I0485725

2012 TRUTH TRILOGY ANCIENT MYTH

Author: MicroStar & Translation: Vicky Hsieh

Contents

IV Starship struck the Earth and Houyi Shot down Nine Suns

V Extraterrestrial Civilization-the Human Crisis

I

The Universe Origin- the Earth-Moon Legendary

Section 1
The Wedge

More than one hundred million years, the things on the planet are getting more and more complex.

Everyone in the chaos really hopes to find out a future direction.

Indeed, in earthly troubles, it is not easy for humans to calm down and find out their directions.

With memories which have enclosed for a long time gradually opened, we will see the truth lost long as well as the human real direction and future ....

In spare time on a day, MicroStar put down the book, closes her eyes and sat into the quiet...

In the light gradually emerging, MicroStar entered a space where time did not exist. (At the same time, she was opened with the function of the penetration of former lives.)
What is the penetration of former lives? When the function of the penetration of former lives is strong, one may know the rise and fall of the community. When the function of the penetration of former lives is stronger, one may see the regulation of the entire celestial changes. And that is the function of the penetration of former lives. MicroStar (the author's pseudonym)

opened her the penetration of former lives in his practicing Dafa, in which she saw the scenes in ancient times before the immeasurable catastrophe:

(500 years ago, a French seer Nostradamus had prophecy ability, which proved the penetration of former lives. In fact, he could also see the past history.)

Section 2
Great God Pan-Gu Opened the Universe and the Little Universe Was Born

The universe was in a chaotic state. Slowly, the disorder universe and the particles in the universe began to rotate, and a new universe was generated gradually. At the time, MicroStar saw enormous Pan-Gu God entering in it as if a man had been reborn. Thus, the universe became the Pan-Gu God body which humans can see by their naked eyes. However, the universe was so large that nobody can consider the Pan-Gu's body was just the small universe. Who would have thought that Pan-Gu's body was not a human form, but the image of the spherical universe?

The small universe had its border and outside the boundary, there were more identical small universes. Different small universes composed of a greater cosmic object and many large universes further combined into the much larger universe. By analogy from that to the endless, it was found that the universe was so abstruse!

When I scanned the small universe, I found that the universe originally had its mechanism of biochemically from everything that layers of materials combining into layers of heaven and earth. In the rotation process, the universe gave birth to a massive amount of large galaxies, which were composed of numerous smaller galaxies, and these smaller galaxies were composed of much more and smaller galaxies. With the combination of layers, layers generated celestial bodies, until the small Milky Way, the solar system, and the electrons inside the atoms rotate around the nucleus…in the microscopic, the infinite is reached.

All celestial bodies are in rotation, gradually forming a stable state of the universe. The mechanism of the universe is an infinite mystery!

Between celestial bodies, there are the factors producing all life. At the same time when the layers of celestial bodies are generated, like the process of a baby's gestation, the God creates layers of all living, things, God and creatures in layers of celestial bodies, thereby the universe is generated.

It turned out that the thing "great God Pan-Gu opened the universe" is true, and what Pan-Gu opened is the small universe we are living in.

Section 3
The Wisdom in the Milky Way

Then, I saw an incandescent fireball exploding which was larger several hundred trillion times than the sun, hundreds of millions of fire points flying out and rotating around the fireball, brightly and splendidly in the dark infinity in space, lighting up the space in trillion light years. These nebulae whirled like a whirlwind, carrying trillion stars in the nebula torrent.

The center of this enormous nebula whirling endlessly was a glowing white ball with great density. The surrounding nebula torrent was more scattered, which was a huge vortex like mixing the dough by an electric mixer to make a cake. But, it was a giant whirlpool of light and heat, while the solar system was only the little flour in the outermost edge of the whirlpool, the planets and the Earth being tiny dust spots inside this little flour.

In looking it down, this was the so-called "Galaxy", which was a huge circular vortex with its shape like a saucer. In overlooking from above, its rotation is clockwise. Seeing the shape of that light swirling, with four swirling peripheral arms, I suddenly understood the Buddhist 卍 character originally came from here. It is the symbol of the rotating shape of the Milky Way, representing eternity, bright, extremely great wisdom and ability. One cannot comprehend without overlooking the light swirling of the Milky Way on the top of the fantasyland.

I saw countless trillions of nebula swirling lines in the boundless dark fantasyland, and the so-called Milky Way was the smallest one only. The similar nebula swirling lines were everywhere, uprightly, obliquely standing, and lying in the flat, with far more than one thousand colors like the swastika shape. This is the true source of the swastika character, which is the abstract expression symbol of the running galaxies! The dark expanse space in the entire universe is also among the rotational movement, shaping the swastika. Outside the universe, there are more endless cosmics, interconnecting and intercommunicating.

In the space between nebula and nebula, there are numerous weak gas and dust particles, with the fiber-shaped form of life and they are also wise. They constantly evolve, they can feel me, and I can feel them. They are, of course, not the creatures breathing oxygen, and don't necessarily need oxygen like the earth's life. I feel some of them living among hydrogen, and some living among the carbon gas. I saw countless planets and nebula with numerous life forms with the strange shape, some existing among the thousands of degrees Celsius, and if the earth people are in this environment, they will immediately be burn into fly ash. However, these life forms survive. They also have the approximate Earth human body, but not exactly the same; they are nothing but non-material. From their points of views, they are real and we are non-material since they can enter our body.

Then, I saw the sun exploding, with many fireballs flying out. The little fireball rotated around the sun, gradually cooling and becoming a series of planets: Jupiter, Saturn, Mercury, and Mars. However, Venus is not the descendant of the sun. It is the last planet accidentally breaking into the solar system from the other side of the Milky Way and captured by the sun.

Then I saw the Earth being originally a fireball, spinning in the Void, and then its surface gradually cooled, forming a thin surface of the crust. How small the mankind are! They are too small to be seen!

In fact, all these were controlled by God. Therefore, God didn't create everything without reasons.

Then, I wanted to take a closer look to the Earth, and the Earth scene instantly moved in front of my eyes as my intention.

I saw the Earth and other planets, like the electrons rotating around the nucleus, rotating around the sun and itself in different distances, high and low angles. The Earth serves as the core of the sun, and red flames emitted out from its surface up to tens of thousands of kilometers, and it also rotates around itself. The Earth is so small, not equal to the size of the sunspot in the sun flames.

I saw there was no moon around the earth then. In the ancient literature, it was recorded that the right eye of Pan-Gu became the moon, so that it appears to be the imagination of the descendants to change the truth of history.

After that, I saw the earth would be destroyed and reorganized every one hundred million years.

The human body needs metabolism. Over a period of time, the old cells are metabolized, and then new cells are generated to replace them. The Pan-Gu body of the universe also needs metabolism and his body's metabolism is the blowing up and restructuring within a certain range of celestial bodies. Our Earth is a particulate in Pan-Gu's body cell, so in accordance with the metabolism regulation, every one hundred million years, it is destroyed and reorganized so that the phase of the Earth's time is one hundred million years.

After the stars in the universe exploded, the residual substances float in the air, forming large amounts of dust and small stars. There are many such floats in the space. God combined these dust and small stars and re-created the Earth. Since the cosmic dust and small bodies used to form the Earth are likely to be the materials left from the Earth or extraterrestrial in the past, after geologists and historians analyzed them, finding that the earth has a history over 3.5 to 4.5 billion years. In fact, this is a misunderstanding.

Section 4
The Moon Created by Ancient Humans

I looked back to identify the source of the moon. It was boundless in the vastness of the universe; existing forever… The vast sky has brilliant stars in large amount, with majestic clouds…

In the universe with a long time, disintegration, reconstruction, explosion and reorganization numerous times occurred in the Earth. It went to the end of the last earth, one hundred million years ago.

At that time, the laws of Creator were widely disseminated in the human world. They were powerful that the moral standards of the entire human reached to the unprecedented and a highly developed human civilization was created. The well-developed science and technology at the time were far more than that in the modern, and many modern people think impossible technology now is commonplace at the time. At that time, science and technology was developed so that humans can manufacture a moon to the sky, and it was when the moon was made.

At that time, in order to solve the trouble brought by the night, people created a moon to light up the dark night sky to reflect sunlight and bring light to the Earth at night. The front side of the moon then was completely polished. In the evening, the moon looked simply bright, being brighter than what it is now, lighting up the darkness as daytime on the earth. The back of the moon was put with some control equipment, precisely because of this, so the front was designed to be always facing the Earth, and the back could not been seen on Earth.

However, the role of the moon is not just lighting, but positively coordinating the Earth's ecological environment. The moonlight lighting which was brighter at night was a supplementary light source, making the plant growth better, and the agricultural harvest was much better than what it is now. The forests on Earth were better than what it is now; following which the weather is much better than what it is now. Now, the coordination capacity of the moon on the earth's ecological environment has been greatly weakened. The food crises may more easily occur today, people endlessly dispute for living space.

The moon is a spacecraft which can travel the universe for a long term. Inside of it is empty and can accommodate a lot of people to live in. The middle of the moon is not only hollow but puts a variety of sophisticated control equipment and precise gear sets. The mechanical precision is far more advanced than the most cutting-edge technology. Such machines can maintain the moon orbiting the Earth normally. Prehistoric people used ultra-high density energy to maintain the rotation and revolution of the moon, so as not to stop rotating to today. Also, the moon can drive the moon for long-distance travel in the universe without having to worry about energy depletion. Such energy also provides living energy for people in the moon, such as the lighting inside the moon to provide light energy needed for crop growth at any time and people's everyday life.

This energy used by the moon can also be used for military purposes. The use for weapons can produce great destructive power which is hard to be imagined. In the combat later, the destructive power of the moon weapon system brought the alien Starfleet trying to invade the Earth a painful disaster, so it is presumable that they had a deep feeling. The moon has a powerful weapon defense system to protect the Earth from invasion and destruction outside the galaxy of evil life.

The construction of the moon is what a huge project! Like aerospace engineering now, many scientists provide intellectual supports for the construction of the moon, engineers in different disciplines are responsible for all aspects of the design and manufacture, and a large number of companies provide the required accessories for the moon. A large number of people are involved in the aerospace engineering.

For modern humans, the moon is so large and heavy. So, how did prehistoric people launch it to the sky?

Section 5
The Moon Warship Roaming in the Interplanetary

It was the era of highly developed psychic abilities, when the prehistoric humans attached great importance to the power of the spirit, often using their own spirits to control substances. They will purify their thoughts to let their own minds very pure so as to strengthen the power of the enlarged energy. The energy then was a transparent crystal ore which was very pure and free of impurities (but not crystal or diamond). The crystal ore had high energy to drive some gigantic objects and even make a city get to the heaven for flight. The flying objects in the sky were not only aircrafts but like the giant ship flying in the sky now. And, this energy was very pure and would not cause environmental pollution after running out so that modern people cannot simply imagine.

The moon was made to be hollow by prehistoric humans, so the weight is not as heavy as the solid sphere estimated by modern scientists. After the skeleton of the moon was assembled on Earth, a group of people took advantage of some devices with the high density energy mechanism, meditating on the ground with the idea of strengthening the energy of these crystals ore to allow the huge moon rise into the orbit. Then the subsequent assembly work was done on the track, putting up giant scaffolding, and numerous metal space airships flying back and forth. These airships were single-seat or double-seat. People wore light space suits, constantly working day and night. It looked bustling.

The moon was finally completed. That was also an outstanding exquisite industrial art. The shape of the moon is oval-shaped, with one end always facing the Earth, so from the view of ground, it looks round. But its shell can be opened. When it is opened, it looks like a blossoming lotus. The key opening it was a sword injected with a powerful energy. The strength of its various functions was determined by the level of the driver's nature of mind. A normal person with low nature of mind cannot open it even he gets the sword. Only the cultivators achieving a certain degree of nature of mind can open it.

When I saw here, I could not help feeling surprised at the transcendence of prehistoric human civilization! Even modern people cannot figure it out. It is really unimaginable!

At the time when the civilization ends up (every time civilization exists was determined), human morality did not deteriorate but becomes better than before. This could not destroy it but let it affect the next phase of human civilization, so that humans and the civilization they created were sent to another wonderful space by God. But the moon was stayed.

In the final phase of the last civilization of the last Earth, due to the excessive development of the industry, then air, water, soil, plants, animals, and human food, everything was polluted and mutated. A lot of people had incurable cancer and other strange diseases, and finally they even looked deforming and very ugly.

The environmental pollution problems now we faced are similar to that then, but unfortunately, human beings cannot see the painful lessons in the prehistory and cannot take it as a lesson. Contemporary people pursue fame and fortune, and do not want to learn a lesson from it, so that they will eventually be harmed.

Prehistoric humans also became extremely corrupted, and they had a long-term war in order to compete for energy. Entire Earth humans are corrupted and variated. The Earth has become a garbage ball with rolling karma. God will destruct the entire planet and reconstruct a new Earth to replace it. When the Earth's energy consumed, this period of civilization will be over. Finally, the Earth will explode in the war.

But there were a number of good people saved by an inheritor of Creator on an airship before the Earth exploded, when the airship flew from the Earth, from that a long-term interstellar travel started. Behind them, the wreckage of the Earth was turned into dust and fragments with all sizes pervading in the space and ran around the sun in the form of a debris belt, waiting for God to use them to re-create the Earth

All celestial bodies and planets are given with life and death. "Form, abide, decay and disappear" is a law, forming the existence according to regulation. Everything finally decays, explodes, and destroys, becoming space free materials. Then it will be combined by God into new stars, developing in the last, it is decomposed and formed into new stars...the cycle will come to infinity.

In the moon Interstellar travel, Earth people encountered many alien civilizations, some of which were good, they taking them to the moon, including the Tantric cultivation methods and the Pyramid civilization.

The Tantric practices have different forms from the Exoteric, in which although the practice between men and women can certainly work out, it is despised and not tolerated by the public.

If the practice between men and women widely spreads in the society like the Exoteric, its suspicion of custom corruption will exist. Therefore, God does not allow it casually widely spread in the folk, only allowing it hand down secretly in the tantric temple.

In fact, Tantric comes from another planet, where the ethical standards are different from that on the Earth. In another planet, there is not shame of moral significance to which people on the Earth suppose. It is regarded as a handshake and completely normal.

So there is no moral barrier in the practice between men and women. But on Earth, if the monk goes out with the women raised for the practice between men and women, he will be absolutely blamed by the public. The rulers could not understand this, so in the Hui-Chang-Emperor in the Tang Dynasty, it was certain that Tantra in the Tang Dynasty was exterminated in the Central Plains.

In Tibet, the practice between men and women cannot casually be carried out. Practitioners of the practice between men and women must be above the Stream-Entry Arhat Law realm. They must be soulshifted from the lives of the upper bounds. There must be some relationships of the birthday hours of the both so that they can conduct this practice by the Master esoteric dharma. Before the practice, a seven-to-twenty-one-day religious ritual will be conducted, further purifying their body and mind, with eating special foods and taking drugs. In the Tibetan calendar, at only certain hours can be the practice between men and women conducted.

Like Exoteric, Tantrism needs lifetimes practice if the practice cannot be completed in a man's lifetime. After Lama passes away and reincarnates, the monk with supernatural powers knows his reincarnation orientation through functions. After finding him, he will be taken to the temple to continue practicing, namely "reincarnation." This ceremony searching for the reincarnation ensures that the reincarnation of Lama will practice lifetimes in the same Dharma until the last cultivation, which can effectively avoid the dangers lost from the world that may be faced by Lama in the reincarnation process. This ritual hands in generation after generation, becoming a major feature of Tibetan Buddhism.

The pyramid civilization came from Sirius. When the Earth's people took the moon to the Sirius and exchanged with the Sirius, they brought the essence of civilization to the moon, and finally back to the Earth. From then, there has been the spread of civilization on Earth. And then in a period of civilization, the inheritors of the civilization constructed the pyramid with Sirius characteristics, and made the small hole at the top aligned with Sirius to commemorate the originated ground of their civilization.

These were good alien civilization.

However, there were also bad alien civilizations. Some aliens saw the moon so subtle that they wanted to seize the moon, but they were hit hard by the moon.

This was the first time when the moon showed its military defense capabilities, which were so powerful! In the future since then, it would show stronger destructive capabilities in the interstellar war against aggression. The fate established with aliens later became the excuse why the God in the old universe allowed aliens come to the Earth.

With the trace of the universe, leaving a trace in the interstellar, the moon became the homeless distant wanderer. When will the adherents of the earth go home? However, the home has been destroyed. So, they temporarily stayed in the moon generations to generations...When I saw it, I cannot help feeling a little sad! Although the explosion of the Earth is set, it is sad that the home has been destroyed, people die, and the survivors were forced to leave their home!

Section 6
God Jehovah with Subtle Wisdom Remade the Earth

One hundred million years ago, a huge Western god flew to the solar system.

He had the image of Jewish and the Southern Europe white. He had a majestic stature, high nose and deep eyes, dressed in the clothing of the ancient Westerners, exuding a glory of universal love, solemn and auspicious, awe-inspiring. I felt surprised and said, Jehovah!

The level of Jehovah is even higher than Moses, Jesus, Laozi, Sakyamuni (they are in the Tathagata level) and is the God with a higher level. The purpose of his trip was to re-create the earth. Then, in the solar system, many dust and fragments left after the Earth explosion floated, some of which were small, some like a big rock, some being a few square kilometers, and the largest even up to hundreds of square kilometers. Jehovah used supernatural powers to come together these cosmic dust fragments. In the original, the energy of the Earth has been depleted. In recycling the Earth, new energy must be buried in the earth so that new human beings can survive and develop. Then, Jehovah moved a lot of energy minerals from the universe by his supernatural powers and combined them with the new Earth.

Then, Jehovah commanded supernatural powers with his idea. Under supernatural powers, the combined group of substances suddenly spun up, turning faster and faster, and almost becoming a fireball. Then the rotation gradually slows down and returns to the normal speed. The surface of the new created sphere is gradually cooled to form a thin crust surface. The Earth was formed.

At the time, Buddha and God in the heaven of the universe on different levels were watching. The upper Buddha on the very high level dropped tears with compassion, which went through time and space and fell to the Earth's surface, to the low surface of the sea, becoming the ocean. Thus, the vast sea of the Earth's surface was formed. At first, the story "tears dropping into the sea" of Lord Buddha in the Oriental Buddhist legend is true. If you take the people's tears and seawater together to test chemically, you will find that the chemical compositions of tears and seawater are very similar.

In the sky, gods of thunder and lightning storm appeared, falling throughout the showers, and after completing all that, the gods departed.

When these little mud balls were made, since the Earth was made to arrange and produce the life environment and many gods coordinated, the ground with originally no life began to appear green plants and life, gradually growing the forests and there being beasts.

At this point, there were a lot of gods going to the Earth's sky to investigate. They just took a look in the air since they felt the ground was dirty. Then, there were many gods, and they came in the interval, so the Earth was very lively.

The gods looked at the Earth to see which place was better and which land or mountain was what they had seen in the upper bound, which was taken to the Earth since the upper bound was disintegrated. At the time, in the sky, there were many gods with the glow.

Section 7
The Legend that God Made Humans Is Real. The Gods Include the Yellow, White, and Black Races

After Jehovah made our earth, a part of Gods were sent to the universe and they created humans in accordance with their own look. There were several Gods being sent down. Nuwa made a part of the yellow race and there were other parts that the Eastern made. The story that Nuwa made humans is real.

In Volume 78, The Custom Links in Taipingyulan, it was said that in the generally speaking of the epoch-making, there was no man. Nuwa used loess to make humans. Since this work was too busy and slow, Nuwa could not afford it so she immersed the rattan in the mud, and after drawing out, the falling mud became humans. By this way, humans could be made quickly.

MicroStar saw that the God made man with supernatural powers, not with hands. The body structure is quite complex, so by using stupid ways of humans was not enough. By using the power of God, that work was completed from the microscopic to the most surface at the same time. Also, God did not complete it in the human time but in the time beyond human space. So humans were made soon.

Some people say, in the Oriental legend that Nuwa pinched loess made man. In the western Bible, it is said that God created man by clay. However, our body is not made by clay. In fact, in the eyes of God, the world composed by molecules is made up of soil, which

is dirty. The human body is constituted by the molecular, and of course, it is the soil. Even the air is dirt and people are drilled in the mud. So, God made man with dirt, which in fact, refers to the fact that God made man with molecules. Descendants added a lot of people's imagination in the process of the legend, while deleting the part that people cannot understand. Finally, the legend is far from the truth, only becoming a fairy tale.

I saw four white Gods made four different whites. Jehovah made the Jews, including part whites in the southern Europe. Part whites in the northern Europe were made by other Gods. Arabs in the past were also made by a single white God. Before Genghis Khan occupied the Arab regions, the Arabs skins and images were completely the same as the white Europeans now, but with dark hairs and eyes. Genghis Khan's Mongol armies captured there and then mixed with the locals. So now their skin is much like the Chinese people, with both the images of the Chinese people and the characteristics of Europeans. Judaism, Christianity, and Islam are the three major religions of the whites. The three major religions came in different periods of history, with the history, the relationship and recognition to one another since they are from the same system in the white world.

Blacks were made by the different Black Gods. Since they were made by different Black Gods, their images have differences in between.

Indians were made by Indian Buddha so that the dance, mannerisms and posture of the hands are like the fingerprints of Buddha. Indians are really a nation made by Buddha.

Egyptians were made by the Egyptian God. The ancient Egyptians were really the red men who may not be able to find now since they were mixed by the Black.

In addition, Persians were made by the Persian god. The Persian nation was a separate nation and also created a brilliant culture.

I saw different gods made the people of different races according to their own images. Asians were made by Asian Gods, Westerners by the Western Gods, Blacks by the Black Gods, other races by the Gods of the same image with the races.

The myth that Gods made men turned out to be true, that was how the humans today came.

Since then, human live and breed in different regions on Earth, gradually from barbarism and ignorance to civilization, ultimately creating a different splendid culture.

II
Guarding the Earth and the Moon Civilization

Section 8
Protecting the Earth's Ancient Humans of the Moon and the Five Kingdoms under the Ground

After the birth of a new earth, the inheritors of the Creator knew that their new home planet has been made by God. So they started the moon and flew to the solar system, stopping the moon in the orbit which was very close to the Earth (closer than the distance between the moon and the Earth now). Since the moon looked like a spectacular mountain which did not connect with the ground, so the modern ancients said the moon as the "ill-Hill."

The interstellar travelers had come from the distant lands, when several generations of time had passed. Since the people then rescued to the moon have reproduced for many generations, the space in the moon had gradually seemed crowded. So they went back to the Earth.

But God did not allow them to live in the ground, because the ground had had a new creation of human. God had arranged their developing for new human beings, so God did not allow them to interfere with the people in the primitive society of the new Earth so that they were arranged to go to the places within the world.00

I saw there was a deep hole without ends respectively in the north and south poles of the Earth, with the shapes like the interlinked beads. The earth's crust of the center of the Earth is the continuation of the crust of the Earth's surface. The crust winds along the hole, wrapping the mantle and the becoming a hollow internal. The topography of the external crust is very similar with that of the internal crust, with continents, oceans, mountains, rivers and lakes. However, only the internal world faces the core of the Earth.

Geographically, the Earth is a solid sphere, composed of crust, mantle and core. We live on the surface of the earth-surrounded by a mantle crust. However, the earth structure that I see is completely different, with a hollow Earth and people live in the internal world.

The way of life of the people living in the internal world is very special. Because there is no sun, the light in the internal world is made by humans. The light source is the five huge spars hanging in the sky of the internal world, which are blessed by many practitioners. After the moon absorbs the positive energy in the moon space, the energy is transmitted to many mountains on Earth (such as China's Five Sacred Mountains). The high mountains transmit the absorbed energy to the five huge spars hanging in the internal world as antennas. The spar energy protects the underground world with sufficient light and flourished bounty. The light in the underground world is more diffuse and moderate than the sunlight. Therefore, the underground world has lush plants, many animals, and adequate pure water. But now, since the humans in the earth surface destruct mountains and lakes, the life of the underground people becomes extremely difficult.

The law of the underground world is very different from that of the human society. There, the biggest crime is to profane and libel deities. The criminal who commits the guilty would be put to death and his body cannot be buried but thrown into the lava lake to destroy. That is because the underground people think the guilty has great karma so that the criminal may be filled with corrupt substances and it may cause great environmental pollution if bury.

I see the Moon Men back from the interstellar travel set up five countries in the underground world. Sine have the most advanced technology, they often drive the flying vehicle, moving between the earth and the moon in the solar system. Also, they are the defenders of the Earth. If there are the aliens invading, they will resist and defend the common homeland inside and outside of the Earth.

Due to their high morality, the underground people's life is both rich and happy. They are generally very longevity. Especially, their kings are the practitioners, particularly long-lived. When they practice to a certain level, they will fly to the moon to live.

The underground people are the friends we can trust! Although they do not contact with the ground people, but they would sometimes imply and teach the ground people with some knowledge. (In the Tibetan legend, the Kingdom of Shambhala and the underground people are the same thing.)

Section 9
The Extinction of the Five-meter-high Giant Family and the Dinosaurs Sixty-seven Million Years Ago

About 65 million years ago, the earth was not rotating around the orbit now, but near the sun closer. On the Earth, it was always in high temperature in a year and the annual mean surface temperature reached above 45℃. Then, humans lived mostly in caves or crypt and the continental plate was not like what it is today.

There were two major continental plates in the Earth's northern hemisphere, inhabited by a giant family. In the Southern Hemisphere, the middle and small humans lived in four continental plates. Between the northern and southern hemispheres, there were some giant islands and archipelagos, with giants, middle and small humans living together.

The original humans made by Gods included the giants, middle and small humans. The giants' height was five meters in average. The middle humans are the people like us, with the average height less than two meters. The small humans' height was only a few inches. The big animals like dinosaurs were prepared for the giants. The five-meter-high people saw the dinosaurs just as we see cattle and sheep.

There is a huge difference between the animals and plants on the land and those on Earth now. Various types of dinosaur populations multiplied on the Earth's land, in the sky and ocean

65 million years ago. The plants on the land were mostly tropical broad-leaved trees and algae wood, when the earth was a dinosaur park.

In a calm morning, the celestial bodies changed inadvertently. The moon suddenly left the track and raise up, flying resolutely outside from the ground. The moon left away from the earth, flying outside of the solar system.

The people on the ground did not notice this change but still live step by step. The pterosaur unfolded the nearly-ten-meter wings, chirping toward its prey with a shrill tweet in the sky. In the tropical broad-leaved forest, a stegosaurus took its young son quietly drank water at the water's edge. In the not far, a Tyrannosaurus Rex greedily stared at the stegosaurus mother and son.

Suddenly, a very special, great earth-shattering sound came in the sky. All dinosaurs in the forest pricked up their ears toward the direction of the sound. The people stunned and raise their heads, looking to the sky in the distance. An asteroid with a diameter of 25 km (as a medium-sized city size) crashed into the Earth in the speed of 30 kilometers per second, with the huge roar from the air friction. The thrilling sights made all people astonished; they looked at the sky blankly and seemed not to realize the seriousness of the situation.

Shortly came the loud noise and the entire planet was in trembling. The Earth was unexpectedly knocked away from the original orbit and the sun. After the asteroid hit the Earth, the generated destructive power was dozens of times than that of Hiroshima explosive. With the asteroid impact, a few pieces of continental plates sank immediately. The rock slivers, dust, and the vaporized seawater surging after impact form a huge mushroom cloud, rising continuously to cover the Earth's surface and obscure the sun. The Earth plunged into darkness. At the same time, the huge tsunami up to five kilometers swept the world. The followed torrential rain continued for several days. Even more frightening is the asteroid impact, triggering the upheaval of the crust and continental plates. Some terrestrial island sank into the ocean, and some undersea continents surged out of the sea and stabbed into the sky with the asteroid impact force. The once ocean becomes the land, and the once land sank into the deep sea with all life and civilization. Since the earth's crust and the continental plates were unstable and moved constantly, the countless volcanoes of various sizes caused by the impact erupted constantly so that the Earth was filled with smoke. The heavy leaping volcanic ash covered the Earth's surface, with this situation lasting for 32 years.

Since there is no sufficient sunshine and plants and animals almost died on Earth, dinosaur died gradually and the earth entered the ice age. Humans of this stage ended and the remnants of humanity entered the next phase of the human civilization period.

In fact, all of this was from God's arrangement. The giants consumed too much of the material resources of the Earth, and the proportion with the Earth was not coordinated. The giant's body was so big that the space distance was relatively shorter and the time was relatively faster. Therefore, the Giants were not suitable for this earth. Thus, in the history, Gods began to phase out the Giants and some Giants went to another space. Animals and plants were prepared for humans. Giants were eliminated so that the megafauna dinosaur prepared for them were naturally eliminated. Therefore, the dinosaurs would become extinct and this thrilling catastrophe came.

Later, the moon returned back to the near-Earth orbit.

Section 10
The Interstellar Navigation Ruins-the Pyramid Civilization

Major disasters made me shocked. When I wanted to heal the mood, suddenly the desert camel shadows emerged and the pyramid scene flashed into my eyes. I saw the historical origins of the pyramid civilization.

Historically, the secrets of the Great Pyramid were covered in a very long period of civilization of the mankind and there was no one who understood the origin and function. I saw that was the historical period long before the human civilization, when the people were five meters high as the giants. Their technology had developed to a high degree in the process of the development of civilization and the field of aerospace technology. To launch a manned spacecraft into outer for space travel expedition, a navigation device was needed. The pyramid was the aircraft navigation equipment built by the giants at the time to explore the universe.

The areas of the pyramid then were not a desert, but a paradise surrounded by mountains and shaded by trees and flowers, where people were happy to live. Why did the small holes at the top of the pyramid face the Sirius? That is because their civilization originated from the Sirius. In order to commemorate their cradle of civilization, they pointed the top of the pyramid at Sirius when building the pyramids.

The pyramid was so huge and the building technology was so extremely

superb that we may not build it with the current level of technology. Then, how was the pyramid built? In fact, we can build skyscrapers. The giants were five meters tall so that they moved large stones much easier than what we think. Plus, the prevailing technology was much higher than what it is now. The mechanical equipment they used was much stronger than that we used now, so as the weight, so that it was very difficult to build up.

After the pyramid was constructed and a very long time passed, the moral of the people had extremely corrupted. Then the continental plates sank to the bottom of the ocean and the original human civilization disappeared.

Later, due to the movement of the continental plates, the plate here raise up again from the seabed. With a lot of sand, plus the tropical and windy climate of this region, a desert was gradually formed.

To create civilization here, God opened up a river, the Nile. Later in this region, new residents were reproduced and people came to realize the efficacy of the pyramid, the long-term preservation, which people knew inside of the pyramid, the corpses can be well kept.

Later, the ancient Egyptians copied a lot of small pyramids, in which they put the bodies inside. However, after thousands of years in wind and sand invasion, most of them have collapsed, but only the Great Pyramid is still standing in the vast sea of sand, staying on accompanied by the camels over thousand years.

Since there are old and new, big and small pyramids, scientists cannot understand how and when they were built, and that becomes a mystery.

I see during the billion years, countless changes happened in the geological structure and the oceans and land lifted ups and down numerous times. There have been numerical creatures on the surface of the earth and new species appear in almost every change. With the changes of the earth's crust, countless civilizations were born and disappeared. Human civilization has been appeared for numerous times, from a few thousand years to tens of thousands of years, one after another in succession.

Section 11
The Frozen Antarctic Was the Treasure Trove for the People to Come

I see a huge land suddenly burst open, and eventually split into two continents, leaving away to different directions. I carefully review that the two continents which split off are exactly the African continent and the South American continent since the Atlantic coastlines are consistent between the two continents so far.

Then, the Eurasia continent drifted eastward and Africa also to the northeast, when both contacted to surround the Mediterranean.

The Indian continental crust was originally under the southeast of Africa, gradually moving to the northeast and hitting Tibet plains in southwest of Asia. The collision led to the rising of the stratum, making the highest peaks there push out above the sea over eight thousand meters, with the land higher than the sea level for few kilometers, ultimately forming Tibet Plateau and Himalayas. We call the highest peak as Mount Everest.

In a flash of thought, I see another scene. Near current Sumatra Island in Indonesia, the Indian Ocean in the south of the equator line was once the location of Antarctica.

The Antarctic is a frozen world now, where the wind is great and there is almost no plant in addition to lichens and mosses growing in the continental margin. There are penguins, very rich mineral resources, and scientific research stations set since the middle of the last century. At the time, Antarctica was located in the subtropical, where was a warm continent. The civilization in this continent used to be much more developed and its technology was better than us. But with the development to the late period, since the moral corrupted, its fortunes began to decline. The last king of the continent's largest country did not believe God and was very selfish and demanding towards the people there. The king did something disrespecting God and he did not know there would be a big disaster approaching.

After the king published remarks disrespectful of God, earth shattering suddenly happened! Earth began vigorously swung, a strong earthquake sparked waves of tens of meters, crazily attacking everything. Moreover, more than a dozen volcanoes collectively erupted on the continent, volcano ashes filled in the sky. Soon, civilization over ten thousand years forever disappeared under the natural anger. Human had no defense force in the face of natural power.

The continent began to rapidly move southwards into the Antarctic boreal. At the time, outside of the Earth, a lot of minerals falling were buried in the continental interior, making the Antarctica become a "cornucopia" with very rich mineral resources. According to the survey, Antarctica has rich reserves of mineral resources, in which there were more than 220 kinds, including coal, iron, copper, lead, zinc, aluminum, gold, silver, graphite, diamond, oil…etc. Also, there were thorium, plutonium, uranium and other rare minerals with important strategy value. Scientists estimate in the Ross

Sea, Weddell Sea, and Bellingshausen Sea, there were 15 billion barrels of oil and 3 trillion cubic meters of natural gas. The coal reserves in Antarctica were about 500 billion tons. From the south of the Victoria Land in southeastern Antarctica, there were extremely rich reserves of coal, with the coalfield area reaching 20,000 square kilometers. This is intentional arrangement of God, because it was reserved for future human use, and people were not allowed to develop here, it must be sealed to hide.

The extreme weather conditions in Antarctica was just used to protect the mineral resources, and God intended to make it colder, windier, and completely frozen, that was why Antarctic was colder than the Arctic. Antarctica is the coldest place in the world, called the "cold pole". The average temperature near the South Pole reaches -49 ° C, in the cold season reaching -80 ° C. Such cold weather is a terrible threat for humans and all life. In the Antarctic, disabilities due to cold and frostbite occur frequently. 98% of Antarctica is covered with ice, with the ice thickness up to hundreds and thousands of kilometers, with an average thickness of 2000 m, the deepest thickness of 4800 m. Due to the cold climate, in the ice-forming process, there is no melting in the Antarctic. Even if humans find Antarctica and know some of its resources, they cannot develop there due to the thick ice cover and homicide wind invasion.

But God left those resources for future generations, instead of letting them not to be developed by people forever. By chance in the future, humans can naturally discover these treasures so that future humans will also develop a very high civilization.

The moon rotates towards the Earth. Time passes by in the eternal.

Section 12
Atlantis Civilization and Depravity

Humans reproduce from generation to generation, endlessly, the same as the underground people. Many kings in the underground fly to the moon to live as the defenders of the moon at the same time. From generation to generation, the moon has been protected.

Until more than ten thousand years ago, in the manipulation of the old God of the universe, the Earth-Moon system had undergone a huge change. The cause was there was a war between two countries, bringing calamity to the moon.

The country attacking the moon was the legendary Atlantis, which were called as the Gonggong in the Chinese ancient books and described as having people face and snake body (which was just a description as "sturdy frame"), red hair (as the Westerners blond), ferocious temperament and being bloodthirsty. In "Mandarin: Zulu," it remarked: "Gonggong once dominated nine states as a powerful leader."

Indeed, Atlantis Empire was once very powerful and as the king of the world. It was a powerful maritime empire, having huge maritime fleets, and once invaded Europe, conquering many European countries. Then, Athenian polis had a perfect ancient Greek city-state in war and other organizations. Under the command of their king, Atlantis invaded the European continent and beat other cities. But the ancient Greeks bravely stood out and fought with the Atlantis alone. The situation was extremely dangerous, but the ancient Greeks were chivalrous and outstanding, and eventually defeated the invaders, obtaining the victory. They saved a lot of people and generously liberated the people who were the slavery out of the conquest of the Atlantis in the Straits. It was also recorded in the Europe's ancient literature.

Then, where was the Atlantis continent? In the Atlantic Ocean, between the eastern America and the Mediterranean, there is a continent, i.e. the Atlantic Continent, also known as the Atlantic Island. This continent starts in the west from the Gulf of Mexico and ends in the east to the west side of the Strait of Gibraltar, where is a vast continent.

Atlantis civilization was far more than the world today, once the most developed place of Western civilization. Some people think that our civilization is the most developed in the history of mankind. However, the truth is opposite that the human civilization was rare low. Compared with the civilization of Atlantis, our civilization is simply not civilized.

Atlantis imposed a transparent mineral crystal, Tuaoi Stone, to provide energy. Tuaoi Stone is a high-density energy, transparent like crystal and diamonds (but by no means is a crystal or diamond). I saw the Energy Center of the Atlantis, where was a vaulted building in the city, including a spacious room and the floor paved with sandstone brick. In the middle of the room, there was a huge crystal-like Tuaoi Stone, placed in a round box on the black base and its role was to supply energy for urban.

Atlantis had highly-developed technology and had made great achievements in the aerospace field. They had the aircraft similar to UFO for long-distance travel. They used to take a special fuel-powered spacecraft to fly to other planets to carve portrait on the rock there and left the relic where they had been.

The Atlantis developed extraordinary ability to travel their body by thought. Surprisingly, the trip was not just confined to the three-dimensional space. They could move their bodies from a place to another in the universe. For example, if they wanted to go somewhere, they only needed to close their eyes and focused their concentration to that place, and then there would be a slight buzz. When they opened their eyes, they would be already there. That was really wonderful!

Atlantis had lot of information received telepathically by the wise. They had a special ability to receive information, similar to the satellite receiving station. They were very accurate, and their work was just to sit there receiving the information coming from other places. That was such a life full of spirits!

In this civilization, the technology was very advanced and there was no serious disease due to the developed treatment. The used treatment method was different from that we use today, which was a combination and use of Tuaoi Stone, music, color, aroma and herbal treatment to play the complete treatment efficacy. The Atlantis can usually live about 200 years old.

However, when they were developed to the late period, the Atlantis people's moral had been largely declined. Jealousy among people was intense. In order to satisfy the greed, they did their utmost and became very dissolute and fallen. Atlantis had no the institution of marriage so that a lot of people were promiscuous on grounds of "freedom." Their sexual life was erosion and confusing.

The Atlantis overly stressed respect for the individual and did not suppress and punish many variation behaviors seriously damaging the moral. Many wise men warned of the consequences of these acts, but most people turned a deaf ear to such predictions. So some people had sex with animals or the human -animal hybrids. Then, they could successfully bio-hybrid so that, for example, a horse would have a person's head, which was so scared! Even if the technology was developed, they lost soul constraints so that they would be indulging. And they thought they were better than God in ability and no longer respected God, becoming very arrogant and making a lot of campaigns.

At that time, the doomsday predictions were widely spread, because some people know that Atlantis had come to an end. But most people chose to ignore it, or not interested. Many people left Atlantis to find the New World. Some people reached a faraway place like Egypt.

Section 13
The Battle between Atlantis and Mu Continent, Bringing Disaster to the Mount Bu Zhou in the Moon

Before the major disaster, some of them crossed the Atlantic Ocean westwards to the America. Then America was a place under the jurisdiction of China, lived by the yellow race of the Chinese people. Since the God Atlantis worshiped was Poseidon, whom was called the Water God Gonggong, so they were called Gonggong.

In Records of the Historian: The Complement Century of Three Sovereigns, it was recorded that in the late period of Nuwa, the feudatories included Gonggong. When they had the territory, their ambition to dominate China arose, and they no longer obeyed China. So, a war of aggression toward China was launched.

Chinese people in the Americas also developed a high civilization and their military technology was also developed. That was a bloody war which was more high-tech than the modern now. The war was fierce. However, since the technology of the Chinese people in America was worse than that of Atlantis, in the early days of the war, the Chinese people in the Americas retreated and had to resort to Vulca of Mu continent then in the South Pacific.

Vulcan', named Chidi, was revered as the god of fire by the descendants. In Guangdong, he was revered as the God of Nanhai, because Vulcan's accommodation was the end of the south. In

Classic of Mountains and Seas: Overseas South, it was recorded that Vulcan in the south had a beast body and human face, who always rode two dragons. Of course, this was hyperbole.

Vulcan was called for the people in the Mu continent in South Pacific, since they lived in a huge continent called Mu in the South Pacific. The continent accounted for much of the South Pacific: it started from Tahiti, in the north to Hawaii, east to Easter Island, and west ended Mariana Islands. It was about 7000 km and the width was about 5000 km from the south to the north. The residents living on this continent included yellow, white, and black complexions of the races, without distinction, and lived in harmony. Then, the Mu continent was a vast empire. They worshiped Titan, the god of fire.

Mu continent's science and civilization were developed. The Mu civilization developed a unique light energy civilization. In that era, the scientific study of light energy and religion was very progressive, and everyone knew how to improve his own light energy as the main practice.

Vulcan emphasized on solar energy, and gave solar energy two meanings: One regarded light as a sacred thing, showing the glory of God. The other identified light as something useful. At that time people used the huge solar energy increase device to convert solar energy into a powerful energy for lighting, power, military and aerospace fields. Mu continent had entered the era believing the sun and its science.

When the allies asked them for help, Vulcan came forward. Then the Vulcan king fought back the invaders at the northeast based on the southwest base camp. Gonggong believed the water god and Vulcan believed the fire god so that the war turned out to be "incompatible."

The war was arduous, particularly in the extremely intense final battle. In this war, both parties sent high-end weapons which were capable of aerospace flight. The Vulcan's fighter fleets were able to send a powerful laser array, so that wherever the laser went, enemy bombers were all ruined. Many frames fell with black smoke dust, and fire arouse everywhere on the ground. The Atlantis people were not the vegetarian, enormous energy emitted by the Tuaoi Stone weapons led to Vulcan's losses. But Vulcan's win was for its volume, many times greater than the enemy firepower. After the desperate combat, both sides had an intense fight.

By Vulcan's forge kills, finally in the use of solar energy consigned to the flames, the air forces of Gonggong were burned and completely annihilated, and the rest had fled back to the base.

The Vulcan king ordered the whole army up the victory and track to the Gonggong's command base in the Americas. After some horrendous bombardment, the rest was overturned. Vulcan asked them as surrender. However, the Gonggong king did not agree and led the residual machines flee to the east. Vulcan pursued and attacked in the full line, from the low-altitude mountain to the peak so that Gonggong had nowhere to go.

The residual army of Gonggong had to angrily counterattack. But under the powerful laser war made by Vulcan, the fleets of Gonggong were destroyed and finally only the flagship fighter of the Gonggong king was left. Vulcan again asked Gonggong to surrender.

The Gonggong king had been in dire straits, then really furious. What could he do? Surrender? Absolutely not! This degraded the great glory of Atlantis. Do impact? Before the opponent became the casualty, they had been destroyed by their laser-war boom. Flee! The flagship fleets and warplanes flied to the sky. At the time the Gonggong king was immensely grief and indignation! He thought that with such a powerful technology as he had, they were soundly defeated. How could not he feel reconciled! On the occasion full of angry, the Gonggong king looked at the sky and saw the distant horizon with the waning moon.

A terrible thought suddenly flashed in his mind, "if you would not let us live, I would not let you feel better! The killed brothers, don't be sad, I would do a major blow to the opponent before I die." The Gonggong king decided to order to crash into the moon with full steam.

The flagship warplanes roared straight on cloud empty, and Vulcan's air force pursued to keep up with them. The Gonggong's aircraft crashed into the sky with a shiny fire, roaring the moon with a power.

When Gonggong's aerospace workers were getting closer to the moon, Vulcan Air Force felt the opponent would launch suicide attacks to hit the moon. Vulcan Air Force hastily pulled the control lever and flew away.

In the critical juncture, a loud noise came from the blue sky. A huge fireball raise from the lunar surface, and the moon violently tremored. The high-speed crash and the huge explosion bombed the lunar surface. The Gonggong's aerospace ship were seriously destroyed. Scattered debris have fallen strongly, burning through the Earth's atmosphere as red as blood, mapping the sky red. The people on the ground saw the blood billowing down from the sky, followed by the fire on the ground.

This is the story of the legendary "Gonggong infuriated the Mount Bu Zhou," in which the moon was the legendary Mount Bu Zhou. In Classic of Mountains and Seas: Wild West Classics," it was recorded that there was an incomplete mountain in the northwest corner, called Mount Bu Zhou. At the time the moon ran in the near-Earth orbit and close to the ground, as hanging in the air. So as the old saying, there was a mountain without linking, named Bu Zhou. Considering if the moon was really knocked down, it would have definitely been a human disaster.

Section 14
The Punishment from the Gods. Two Continents Sank into the Bottom of the Sea Overnight

The moon shell was damaged. The two countries waging this war have been severely punished by God and their homeland and the people were sunk.

Ten thousand years ago, the continent of Atlantis sank. In Currie Iglesias, Plato recorded the conditions before the Atlantis sank: "In the Atlantic far away from the distant western Mediterranean, there was an amazing continent. It was decorated with gold and silver and produced a gleaming metal. It had well-equipped ports and vessels, as well as the aircraft which can convey people for the flight. Its influence reached the distant African continent.

Atlantis, also known as Atlantis, as the country called the Atlantic States, was a huge island in the Atlantic Ocean. In the island coast, there were many mountains, with an open great plain in the middle. The island was rich with minerals and a variety of animals and plants. The capital city was located in the center of the island and a bustling metropolis. In the town center, there was a royal palace and the temple dedicated to Poseidon. The palace was magnificent and lavishly decorated. Three broad canals surrounded the main island and the entire island was divided into a number of concentric circle areas. Another canal was throughout the district from the center to the coast. Atlantis experienced ten emperors and at the beginning, it was ruled with virtue and had been very strong and a prosperous country. It always had pilgrims and businessmen coming and been bustling.

However, with the abundance of life, their holy had gradually disappeared and the Atlantis community began to become degenerate. They were snobbish and struggle for power and money. They were addicted to desires, abandoned due virtues, and liked militancy and tyranny, so they frequently launched war of aggression, one after another conquering Europe, the Mediterranean and Egypt. Therefore, Zeus decided to punish the Atlantis. He summoned the gods to his temple and said.... Plato's record came to a halt.

Since some Atlantis had superior psychic ability to know in advance the end would come in the near future. In the last few weeks before the sinking of the mainland, they protected the Tuaoi Stone with a transparent cover magnetic from damage in sinking. One day, when it is found again, it will prove the existence of the civilization of Atlantis and may be used again.

In a sunny noon, a strong lifting movement of the plates started undersea. The violent collision of the earth plates triggered a major earthquake. Buildings bumped and shocked up and down. At the same time, Earth magma, of the volcano, erupted. The splashed Mars caused land fire, which made the smoky sky red, blocking out the sun. Land subsidence and earthquake triggered a terrible tsunami, and accompanied by the typhoon, high waves surged over a hundred meters   on the sea. The land was sinking as if the sky had fallen down. Seawater surging shocked and devoured everything, and eventually flooded the earth. The Atlantic was like boiling water tumbled. People fled in all directions, swallowed up by the flood or fell into the fire pits, with endless hiss screaming. Earth was collapsing as if the end of the world had come.

Overnight, the continent of Atlantis sank in the Atlantic, buried in the seabed. All the people on the island died in the belly of the fish and the former glory of everything about this civilization had become a beautiful legend.

The continent of Atlantis sank, causing huge tsunami on both sides of the Atlantic and destroying many civilizations along the coast. Europe, the west of Africa, and the east of Americas were flooded, devastated and collapsing the population.

The land and people in the Mu Continent on the continent of South Pacific also sank to the bottom of the sea.

In fact, although the Mu continent civilization had increasingly showed prosperous and science and technology, culture, economy were highly developed, there had been always the potential danger. In the Mu continent civilization, under prosperous economic life and increasingly low-grade entertainment culture, people's morality rapidly declined and decayed. There was the widening gap between the rich and the poor. Ethics, education, culture and social order problems occurred. Even depraved sexual life, sex trade and drug abuse, trafficking and other criminal acts appeared in the whole society.

Finally one day, disaster suddenly happened, that was a punishment from God!

The shocking tragedies first came from violent volcanic eruption. A terrible roar happened in the Mu continent. All in a sudden, the sky and the ground were broken and volcano erupted. Many wealthy cities were quickly swallowed by lava flows. New volcanoes constantly burst. The pillar of fire and thick smoke covered the sky. The flames made the sky red, as if the end of the world. In the angry sounds of the crust, a strong earthquake and violent tsunami were also triggered. The sky shook and the ground moved. The earth fluctuated and cracked like waves. Some peaks bombed like a nuclear bomb. The mud and stones were rushed to the high altitude and scattered down. The terrorist horror was so great. Everything on the Earth collapsed like blocks. Numerous buildings cracked and fell down. The roads were cracked and collapsed. The communication interrupted. The tragic scene cannot be described by words easily. Soon, the whole continent was ruined and then began to sink in the rumble. Overwhelming tsunami flew indulgently. The entire earth and its towns, forests, people, and animals all sank and countless lives disappeared in an instant.

Only overnight, the Mu continent was drawn to sink into the sea by a huge power and the once brilliant Mu civilization was wiped then. As time goes by, no one remembers there ever had been such an ancient continent gradually.

The Mu continent sank in the seabed and the huge amount of water around was quickly poured into the subsidence area. The people near the Pacific coast could see the seawater suddenly retreat away southeastward like being pumping. So, the ancient people recorded in the then "Trap Southeast." In "Huainanzi: Astronomical Institute," it was recorded that "the ground in the southeast was not complete, and the water and dust left."

However, the suddenly receding of the sea water is by no means a good thing! The people who have experienced the tsunami in South Asia that before the arrival of the tsunami, the sea water anomaly receded, and when it come again, it is buttressed by an avalanche of irresistible force shock waves.

This tsunami was caused by the Mu continent, of which the momentum and harm was far more than South Asia tsunami in hundred to thousand times. Suddenly, the monstrous wave up to a hundred meters roared from the sea, with huge impact overturning and all wrapped. The whole process of tsunami was a scene of terror and death. When people looked up distant waters, they saw the overwhelming impact of oppression from the top and deep blue of the mountains. All the people were scared to scream and ran away. However, in the face of the raging tsunami, how small human beings are! Who could escape from this catastrophe? Under the impact of the tsunami, the Pacific Rim countries immediately became the ocean. In the rolling waves, dead bodies and garbage were everywhere and drifting with the tide.

It was a devastating catastrophe! After the tsunami, the tragic scene was shocking. Countless people died of this monstrous tsunami. Countless dead bodies drifted in the water. Cities were continuously destroyed all

and the ever-rich land became rivers, lakes or wastelands. The whole earth became the ruins. No one knows how many years the survived civilization had gone backwards!

Mankind is so small in the face of the enormous power of the natural and life is instantly swept away by the floods. That scary scene is easily to shock people. Many years later, when I saw back the tsunami, I would be still terrified. The sink of the continents in the East and the West as well as life into the flames in the glowing lava made me sad.

III
The True History of the Ancient Myth

Section 15
Nuwa Repaired the Sky, Assisting the Underground People Repair the Moon

Another scenes flash into my eyes and I saw the old continent's civilization ruins still exist in the seabed around Bermuda and the shallow water in Caribbean. There were the huge artificial stone walls in the seabed, stretching in hundreds of kilometers. I also saw the old sun in the southern Pacific Ocean falling under crust, then rising parts of it, becoming the steeple islands.

Houyi and Chang-Er were once the king and the queen of the country in world under the ground. After they practiced to a certain extent, they flew to the moon and became the defenders of the moon. They practiced the power of life, so their bodies could appear in every space. To their ability, they could have protect the moon and not be hurt. However, due to the catastrophe made by God, they did not dare to disobey.

After "Gonggong infuriated the Mount Bu Zhou," Houyi and Chang-Er quickly raised the moon on the track now and proceeded to repair its shell.

The moon left the near-earth orbit and rose to the track in the sky now. The ancients saw the incomplete mountain originally floating in mid-air disappeared so that they thought the ill-Hill collapsed. As a result, it was reported

in the Records of the Historian: The Complement Century of Three Sovereigns, "Gonggong could not help getting anger so as to touch the ill-Hill with his head. The sky was folded and the land was broken."

The moon left the near-Earth orbit and was gone, and the mutual attraction between the Earth and the Moon became smaller. The tremendous gravity and the pull in separation made the orbit plane of the moon around the Earth deviated from the original one. Both the gravity and the pull made the Earth's axis slightly offset and moved towards the present location in the direction that the moon was leaving. The ancients found that many stars fell down to the northwest in the northern hemisphere night while seeing the sky, thinking that the sky in the northwest tilted. Therefore, it is reported in "Huainanzi: Astronomical Institute, "the sky tilted to the northwest and so the sun, the moon, and the stars moved."

The legend collapse hole in the sky after Gonggong could not help getting anger so as to touch the ill-Hill with his head refers to the fact that the lunar surface was bombed and collapsed.

The world had just experienced a tsunami, and so the civilization was hit and would no longer reach the level as ever before. Thereafter, the people living in the internal world refined stone in the world to obtain special metal materials to patch the moon. Then, they drove the airship several times to send the refined metal blocks to the moon to repair the moon surface after high temperature melting. In the Central Plains then was the late dynasty ruled by the Nuwa (this Nuwa was not the Nuwa God making humans but a clan believing in the Nuwa God). The Nuwa supported the people living in the internal world fully for the moon repairing project. Therefore, it became the story of Nuwa Repaired the Sky. The myth of Nuwa Repaired the Sky came from the "Huainanzi: View Training Offerings," as follows: In the ancient time, the quadrupole wasted and the Jiuzhou cracked. They sky collapsed and the lands could not load. The fire was not extinguished. Waters and oceans were endless. Beasts ate innocent people and birds preyed the elderly and infirm. So, Nuwa refined multicolored stones to fill the heavens off, Then Nuwa cut four legs of a big turtle as four pillars, supporting up half of the sky collapsed. Nuwa also killed the black dragon harming people, removing the arrogance of the dragon. Finally, in order to block overflow flooding, Nuwa collected a lot of reeds and buried them into ashes to block the floods spreading. The sky was still tilted to the northwest. Therefore, the sun, the moon, and stars naturally turned to the west. Because the earth was tilted to the southeast, the rivers ran there. When a rainbow appeared in the sky, it was the intense pulsed light that Nuwa used to repair the sky. Every rainy day, there were many floods which surrounding people suffer unbearable. The people living in the internal world sympathized human disasters so that they refined the stones and sent the gravel and ashes to the lower-lying areas and built high dams to stabilize the flood, by that people can live peacefully. Ancients were gratitude for what they did, and then recorded the matter: The Nuwa's "gathering ashes to end floods."

Even though the moon was repaired, it was not as good as before and all aspects of functions were greatly damaged. In the past, because of the benign and coordination capacity of the moon on the Earth's ecological environment, the mankind were not as bad as the mankind today. As a result of the food crisis and the fight for survival space, a huge negative outcome has been caused, so that the Earth would not prematurely become a karma ball to be destructed. Now, however, because the moon was largely hurt, the ability to adjust the ecological environment has been greatly weakened. Many forests on Earth have gradually disappeared and the climate gradually deteriorated. The core energy began to weaken and the technology owned by the underground civilization is not as precise us the past.

Section 16
I-Ching Knowledge Led to Flourished Publication and Saints Followed

I saw the thousands of years have passed and the human regime has changed. Nuwa, Fu-Xi, and Shen-Nong were followed by replacement and were the Three Sovereigns in the history of China.

The two continents in the East and the West sank, causing a huge tsunami so that human civilization was ruined, followed by a new period of human civilization.

In this case, Nuwa declined and Fu-Xi rose. Each rising of the clan is the rise of a dynasty, not just one person's rise. We study the ancient history and find that every "clan" can last for hundreds of years, some as long as four or five centuries, or even longer. It is impossible for a person to live such a long time. The dynasty of only one generation can continue for so long.

Fu-Xi is the Chinese cultural ancestor, leaving tangible cultural heritage for descendants- I-Ching.

Many modern people may not understand what I-Ching is. "I" means changes. I-Ching is a scholarship to study the variation of the universe movement. The scope of the study covers the movement changes in the entire universe, the entire society, and even everyone's life experience. If the regulation of the movement changes is found, human survival and development will be better guided. This knowledge is big enough! The Chinese

ancients were in research of so much knowledge.

Recorded in Houwoheji, Hou, Shang-shu," Fu-Xi had the world. Drag-on-Horse had figures on their bodies out of the river and Fu-Xi imitated those to paint the Bagua." The River-Painting were the onset of I-Ching knowledge.

Fu-Xi looked at the heaven and the ground and found the natural phenomenon and laws, and saw the morality of gods to classify the feelings of all things. Through perception and observations, Fu-Xi understood the relationship of the natural things between the galaxy and the world, creating the Bagua map of Qian, Hom, Gen, Zhen, Xun, Li, Kun, and Dui, known as Fu-Xi Bagua.

To the Dayu era, there was a fantastic turtle with "Luoshu" on its back emerging from the Luoshui so that Dayu created the "Hongfan Nine Domains" including the thoughts of "Five Elements." Since then, the thoughts of Five Elements spread in the vast land of Shenzhou, affecting all aspects of Chinese culture, such as ancient Chinese medicine, forecast, and physics which have been affected by those thoughts.

Recorded in "Chou-I: Xi-Ci," figures were from the river, and a book was created in Luo, followed by saints. Since then, I-Ching has been learned in the world.

In late Shang Dynasty, King Zhou believed the calumny of Chong Houhu and imprisoned Hou Jichang in Youli. Hou Jichang was at leisure and played musical instruments   everyday. He extended the original Bagua to the hexagrams with total 384 yaos, including the wonderful fortune of heaven and earth and Yin and Yang mechanism. The post king Wu, Fazhou established the Zhou dynasty, and respected Jichang as the "King Zhou Wen." Since the hexagrams were extended by King Zhou Wen and different from the original I-Ching, the I-Ching he extended was added with the name of his dynasty at the name as The Book of Changes, meaning that it is easily to be learned.

Later, it was said that the I-Ching was made by King Zhou Wen, but it is not true. I-Ching is the prehistoric culture and after King Zhou Wen inherited, he changed and re-edited it.

The Book of Changes's content is esoteric, which can not only provide people for divination but help people to read and learn so that Confucius liked to read it. To read The Book of Changes, Confucius grinded the rope ting the letters three times. The Four Books and Five Classics in the Chinese Confucian are The Great Learning, The Doctrine of the Mean, The Analects of Confucius, and The Words of Mencius. The Five Classics are The Book of Changes, The Book of Songs, The Book of Rites, Shang Shu, Chun-Chiu. I-Ching was respected by Confucian as the first of Five Classics.

As knowledge of the laws of motion of matters, I-Ching is very esoteric. Future generations felt its hardness and sighed: "the knowledge of I-Ching us difficult. People change three saints and the world's calendar precedes three ancients." The three saints are Fu Xi, Wen of Zhou, and Confucius. The three ancients are the Fu-Xi ancient times, the medieval of King Wen of Zhou, and the Late Antiquity era of Confucius.

Section 17
The Battle between Huangdi and Chiyou was an Incredibly Hard One between God and Human

In the eight generation of Yandi, the Yandi forces gradually declined, thereby Xuan Yuan Huangdi rose.

Huangdi rose in a war-torn era. "Records of the Historian" reads: "in Xuan Yuan times, Shen-Nong declined. Feudatories invaded each other and abused people, while the Shen-Nong could not attack them. So Xuan Yuan attacked them. Chiyou was the most violent and could not been attacked." That means when the Shen-Nong went to the eighth generation, when Yandi was the king and he failed to control those feudatories in rebellion. So Huangdi rose and attacked the feudatories in rebellion, while only Chiyou did not obey.

Huangdi controlled the chaos and became the overlord, while Yandi was naturally unwilling. In "Records of the Historian," it is recorded that "Yandi wanted to invade princes who surrendered to Xuan Yuan. Xuan Yuan trained his troops and pacified the population. Xuan Yuan taught brown bears, brave troops, and tigers to fight against Yandi in the wild of Banquan. After three wars, Yandi won. Huangdi defeated Yandi in Banquan, and so officially acquired the throne.

After Huangdi won, he ruled with morality and loved the population. Chiyou wreaked havoc on the people and launched rebellion to challenge to Huangdi. Huangdi loved his citizens and did not want to attack Chi You and persuade Chi You to stop. But Chi You did not listen to advice and repeated violation of the border. Huangdi finally sighed: "If I lose the world and Chi You is in charge of the world, my subjects would suffer. If I tolerate him, it is arbitrary. Now he has no virtue and blindly invades us, what I can only do is to punish the unjust!" So Huangdi personally led his troops to fight against Chi You.

In "Records of the Historian," it is recorded that "Chi You rebelled and ignored Huangdi. So Huangdi led his troops to fight against Chi You in Guru. Princes respected Xuan Yuan as the emperor on behalf of Shen-Nong, known as Huangdi."

In fact, there is not only the king inside the ground but also the king on the ground that Huangdi was the emperor of monasticism. So in the legend, it was said that he was the ancestor of Taoism and had a lot of miracles. When his practices completed, he went back by a dragon. I saw the final battle between Huangdi and Chi You with some episodes like the myth of "Investiture of the Gods." It was not a large-scale war and Huangdi's soldiers were less than two thousand. Chiyou's soldiers were over Huangdi's and very mighty without humanity like savages.

One morning, before the start of the war, Huangdi boarded the high platform and told the soldiers: "Even though the enemy number was more than ours and they looked ferocious, I will invite many gods to assist in the fight and we will eventually triumph."

When the war officially began, Huangdi personally beat the drum of war to boost morale and increase confidence. A mass of dense mass of people rushed over murderously. Heaven clouds were rolling, followed by thunder and lightning, but it did not rain. In seeing the situation and the sky, some soldiers were afraid, and others courageously stepped forward against the army of Chi You.

Huangdi raised a flag to the sky and said loudly, "Gods, help me fight against Chi You to relieve the pain of the people!" Then, many Gods dressed in shining golden armor flew down from the sky to help Huangdi's army to fight. As Chi You saw the gods for Huangdi, he also requested a lot of negative gods. This was really an incredibly hard battle.

In the critical time, Huangdi took out a gourd-like instrument prepared as he came to the folk and threw into the air. Those negative gods were immediately received inside the instrument.

In Classic of Mountains and Seas: Wild North Classics," it was recorded that when Chi You attacked Huangdi, Huangdi asked Yinglong attack the wild of Jizhou. Since Yinglong brought water, Chi You asked the water god and the rain god to intensify wind and rain. Huangdi requested the female fairy called Ba assisting in the fight. When the rain ended, she killed Chi Finally, Yinglong captured Chi You and killed him. Yinglong also killed Kuafu and went to the south; therefore, it always rains in the south of China. In Classic of Mountains and Seas: Wild East Classics," it was recorded that in the northeastern of the wild, there is a mountain named the Evil Plow Mounds. Yinglong lived in the southernmost. Since he killed Chi You

and Kuafu, he could not go back to the heaven. Without Yinglong, there was often drought in lower bound. Once in the face of drought, people dress up like Yinglong to get the heavy rain.

At this point, the sky began very fine and many auspicious phenomena appear. At that time, a lot of Huangdi's soldiers were very excited and Huangdi was pleased. Huangdi beat the drum again in withdrawing troops.

Followed by Fu-Xi and Shen-Nong, Xuan Yuan Huangdi dominated the ancient China and unified the various tribes in China for three times. Then, Huangdi made officials and institutions. He ordered the ministers to build palaces and made the law and the calendar. He ordered Changjei create text and created the law of The Six Categories. He talked the pathology with Qibo and wrote "Yellow Emperor." Besides, Huangdi made the currency with the copper of Shoushan and created the system of measures by making Gan-Zhi. Also, he taught people sowing grain, making clothes and vehicles. Then, arithmetic and musical instrument had been invented successively. Due to great national power, stable politics, and cultural progress, ultimately the flourishing situation that "the farmers were not greedy to take land and the fishermen did not compete fishing waters; no one picked up lost articles in the street and the businessmen did not overbid on the market."

At the same time when Huangdi established the prosperity, he opened a brilliant age with wonderful political achievements.

In fact, Fu-Xi, Shen-Nong Emperor

Yan, Huangdi, later Yu and other sages were the forefathers sent down by the God to teach people, so as to create Chinese civilization for nearly thousand years. We should be proud that we are all the descendants of the Fiery Emperor.

Section 18
Kuafu Ran towards the Sun, Becoming the Indian Ancestor in the American Continent

About 4500 BC, there was the war between Huangdi and Chiyou. Kuafu, the descendant of Emperor Yan helped Chi You to fight against Huangdi. Then, Chi You died of defeating in the war. Kuafu, Chi You, and the people fled away. In order to cover the tribe to retreat, Kuafu courageously severed offspring and was killed by Yinglong in the pass.

The second generation of Kuafu continued to lead his tribe to the east for the refuge in the Weihe River region, where, many years later, there were more and more people of the tribe of Kuafu, leading to less and less living space. Therefore, the second generation of Kuafu led his people to migrate to the northeast, the Lake Baikal region. Unfortunately, the second generation of Kuafu died on the way. The next generation of Kuafu led the tribe to the Lake Baikal.

In Classic of Mountains and Seas: Overseas North, it was recorded that "Kuafu chased the sun. He was thirsty and drank in the Wei River. Due to inadequate water, he ran to the north to drink along the great river. However, he failed to be there and die of thirst. He abandoned his stick, becoming Deng Lin." In Classic of Mountains and Seas: Wild North Classics," it was also recorded.

Many years later, the new generation of Kuafu led the clansmen to immigrate to Heilongjiang. After many years living in Heilongjiang, the clansmen of Kuafu continued to go to the east along the Sea of Okhotsk to reach the Kamchatka Peninsula, then across the Bering Strait to Alaska, North America, and then go to the east to reach the West Coast of Canada.

In the long centuries of the migration process, the Kuafu tribe lived for some time and went for another. In Northeast Asia and the northern part of America, many countries of the Kuafu family were established.

After settled in North America, part of the Kuafu tribe continued to the south. Through Central America, they arrived the South American continent, and along the way had also established a number of countries.

At this point, there are tribes of the Kuafu family as Eskimos in the north and the Peruvian in the south. For hundreds of years, the Kuafu tribe moved all the way east, following the sun in their faith and exploiting the boundary in the land of the Americas. To the present, many Native American tribes still believe Titan, which is the religious belief handing down from their ancestors.

This is the story of "Kuafu Ran towards the Sun." The people of the world laugh that "Kuafu wanted to chase the sun without being within their means." However, since Kuafu chased the sun to bring the fire of civilization to the American continent, which is worth all the descendants of Chinese proud.

When the Kuafu family pioneered the Northeast Asia, where the Chi's and Shaohao's Clansmen came succes-

sively. Since the Chi's did not accept Huangdi's ruling, they moved to the North and then to the Northeast Asia. The shaohao's were one of the Taihao (Fu-Xi) generations and came to the Northeast Asia to expand the living space. The Chi's and Shaohao's gradually integrated with the Kuafu's, so that they moved to the east together, entering North America and diverting to the south.

The Chi's moved to the Mississippi River basin, North America, located in Tennessee, Oklahoma, Kansas, Missouri, Mississippi, etc.. Since the history was too long, even the Indians have forgotten who their ancestors have been.

The Shaohao's lived in southern California and northern Mexico and established the Shaohao Xihe State. In Classic of Mountains and Seas: Wild East Classics," it was recorded that "The Shaohao was located in Dahe outside the East China Sea. Zhuanxu, Shaohao's Ru Emperor, abandoned his harps there. There was the Gan Hill, with sweet waters and rivers." Dahe was the Grand Canyon, Colorado, in the United State. The Shaohao country was the Shaohao Xihe State. The sweet waters were the Colorado River, running into the Gulf of California, as the sweet rivers. The Shaohao Xi country is also known as "Tang Valley Fuso," and Tang Valley is Gan-Yuan. Tang Valley Fuso is now Mexico, in ancient times, also known as the Fuso country.

After Huangdi, there were many dynasties, such as Shaohao, Zhuanxu, Di-Ku, Di-Zhi, Di-Yao. The human regime rises and falls, turned over all the time. In two thousand years, there was no major natural disaster and human civilization had developed to a higher level.

Section 19
The Venus Plunder the Earth, Making the Earth Reversed and Causing a Huge Flood

However, when the wheel of history went to 2297 BC, that was the time of the ancient Chinese Emperor Yao, an unprecedented global catastrophe terrible was impending.

With a lightning-like idea, I found myself looking down the Earth's rotation in the air. There was a reflective flashing on the blue ocean on the sphere. White clouds enveloped the Earth and roll roundabout. The Eurasia and the Americas were like a few pieces of mossy rocks, where human beings, civilization, cities, and great constructions were invisible.

Then, celestial bodies changed again. I saw the moon slowly rising and beginning to leave the Earth. It speeds up outside the solar system.

I did not know what happened!

Soon, the moon left away from the Earth.

In this case, in the universe outside our solar system, a planet suddenly crossed the track with a piercing whistling and a tail with a million stars, merged into a giant light across the Earth.

Hundreds of stars flashed, among which the largest swept out of the earth. Planet's suction was like huge waves and made the Earth scroll up, falling outside the orbit after turning. Therefore, the South Pole and the North Pole exchanged, and the East and the West did the same way. The Earth originally rotated from east to west but from west to east since the South Pole and the North Pole exchanged. The flow of those clouds, currents, and everything else also changed...

On seeing the earth, the sky seemed to burst and a large number of aerolites roared toward the Earth. The whole sky was reflected to be red, like blood. Humans experienced a catastrophe that boiling stones fell from the sky and stones were fired. Flames were on the earth with one miserable scene.

One million luminous stars left away, and finally only a superstar was left, wandering and sparkling in the solar system and becoming a major planet with a rotation direction contrary to the rest of the seven planets, namely the Venus!

I could not help marvel with that, at the same time understanding why the planet Venus's rotation direction is different from the others of the same planet system. Since the Venus broke into the solar system later, it is the only giant planet with a reverse rotation and in the view of the sun rising from the east and falling in the east. In ancient China, the Venus was called the "Vesper", where is the abode of the god Taibai-Jinxing.

The Venus invaded the Earth, resulting in the reverse of the Earth's poles, which became a legend in the human world. In western Brazil, there is an ancient legend: "Sky burst and debris smashed down, smashing things on the ground and the creatures of the earth, switching the position of heaven and earth."

The Venus invaded the earth, tumbling the Earth and causing great confusion on the ground. Seawater shook vigorously, stimulating huge tsunami shock waves more than two thousand meters high as if Mount Tai collapsed. The potential raging was monstrous and devoured everything down. When it impacted in overturning coming, people thought that the heaven was lost. Earth's polar ice caps melted and the entire planet became a water polo. Floods were everywhere and only countable several mountain peaks were out of the water. That was absolutely a big tragic era in the human history.

Section 20
Noah's Ark and the Flood Control by Dayu

I saw some small creatures on mountain peaks, being human survivors. In Alps, Europe, there are surviving Europeans, also in the mountains of Africa. In the Ararat Peak, in the junction of the Middle East, Turkey and the Soviet Union, there is a coffin-shaped rectangular ark running aground at intermountain.

"Is that Noah's Ark?" I said with surprise.

When humans finally developed to the last civilized end, the sin of humans was deepened and God decided to once again destroy mankind. However, there were still good men in the world. Noah was a righteous man and God liked him. One day, God said to Noah, "The world is full of evil and I have decided to destroy all living creatures in the world. You have to build an ark, so that you can save your lives. The ark needs to be established with gopher wood, with three hundred cubits long, fifty cubits wide, and thirty cubits high. Above the ark, translucent windows should be made, with a cubit high. The ark's doors need to be open on the side. The ark is divided into upper, middle, and lower layers and each room should be coated with rosin. You must conduct it according to my arrangement. Take your wife, your family, the variety of birds and beasts in pairs and couples to the ark. You have to reserve enough food. After seven days since the completion of the ark, I will make forty-day rain and the flood will drown every living creature on the ground.

Noah led his family to build an ark in accordance with God's command. The passers-by and neighbors all laughed at Noah. Noah persuaded them, "flood is coming soon. Be early preparing." But people simply did not believe that the flood would come. After the completion of the Ark, Noah led his family and various pairs of birds and beasts into the ark, and stored enough food. God closed the wooden doors of the ark for them.

On February 17, it began to rain, day and night without ending, soon flooding rivers and there was an expanse of water on the earth. The flood swept everything…

God surely brought floods upon the earth.

The Ark had been washed from the flood, bumping in the marine.

The storm lasted 40 days and nights. All the earth, trees, houses, human, animal, and mountains with altitude below two thousand meters were submerged. No one survived.

When the sins of the world were washed away, by the flood God stopped the flood. At this time, the flood had been in the proliferation of one hundred and fifty days.

Waters were waning, and Noah's Ark stopped bumpy and began drifting downstream. On July 17, the ark came to rest on the mountains of Ararat.

The wind blew the water, making the water evaporate and the water subsided. On the first day in October, the tops of the mountains became visible.

On the next year, February 27, the flood waters had receded all. God said to Noah, "Now, you can come out now." Noah and his family went out of the ark one by one. They survived in the ark for one year and 10 days and became the survivors of the world. Birds and beasts and all kinds of animals in the ark were also put out and thrived again.

The great flood completely destroyed human civilization on earth, the mountains below 2000 meters above sea were drowned, and only the people two thousand meters above survived. Western civilization was completely destroyed and European civilization was re-developed from nothing.

The big flood ravaged the world, as recorded in the Bible, "the waters were extremely vast, the high hills under the whole heaven were covered, only the people who lived in the mountains survived." Only a few people lived so high and their level of civilization was relatively low. They left down the mountains because civilization was all destroyed. They could only difficultly "start from scratch" from "nothing" to create a new culture.

The descendants of the survivors were just the mountain residents who did not know the text. They had heard of the prehistoric rulers' name but knew little of their performance. Although they were very happy to tell their names to their children and grandchildren, they knew almost nothing about the results for the ancestors in addition to some obscure legends. Because they and their children had been living since the number of generations among the poor, they only concerned their own needs and what they talked about was their own needs so that the stories of these ancient times were forgotten.

Only after people have the necessities of life and a lot of leisure time do the collection of myths and legends and the study of ancient things exist. That is why many ancient names pass down but their performance has not been circulated.

Plato recorded in "Timaeus", Solon and Egyptian priests talked about the history and today. After a very old elderly priest listened to Plato talking about the ancient legend of the Greeks, he said, You Greece people never grow up. You are all children without any old man."

Solon heard it but did not understand. The priest said, "Due to various reasons, the humans in the past had suffered destruction and will have a catastrophe in the future. Among many major catastrophes were from floods and fires and other smaller ones were numerous. Fire, for example, might be a volcanic eruption or the burning sun, shown in the form of long years of drought, which the people who lived in mountainous areas and dry areas might easier suffer than people living in the river or seaside. The Nile River made Egypt not suffer from such a disaster. In addition, when the gods led to flood the earth, the people living in the mountains survive easier. You Greeks living in the plains were washed into the sea by the flood. However, in the Nile Delta, water never washed into the farmland down from high places but went up from lower places. Therefore, the Nile region is lucky and the great savior of the Egyptians undoubtedly.

For this reason, the things Egypt saved were very old. If there was any significant amazing event happened in any area Egyptians knew, it was written down by the Egyptians and stored in the temple. At that time, people of other nationalities might have just begun to enjoy the text and some other things required in civilized life.

After the last major flood, the European civilization almost raided all, just some people who did not understand the text and lack of enlightenment left. Thus, the descendants of Greeks knew nothing in ancient times. After the death of the survivors in the disaster, many generations were not able to write some text to devolve to the descendants."

When Noah's family took the ark to escape the great flood, the Emperor Yao ruled this magical land, China, in the Far East. In a moment, floods raise up to the sky, surrounded mountains and drowned highlands. On days of non-stop rain, waves were everywhere. China had become a vast expanse of water. A large number of people were drowned and starved to death. Surviving people fled to the mountains, with starvation. People were very worried and apprehensive. Sima-Qian simply recorded this great flood in the "Historical Records," "When the time of the Emperor Yao, people suffered from floods, which spread to Huai-Shan and Xiang-Ling. People were very worried." In "Mencius: Tengwen-Gong," it was recorded that, "When the time of Yao, the world was still unknown and the floods flew cross the ground, flooding in the world." "When the time of Yao, water went in the wrong

direction, flooding in China. Snakes and dragons lived here and people had no places to live in. On the top was the nest and on the bottom was the camp hole." In "Classic of River and Mountains" written by Liu-Xiu to Handi in the Western Han Dynasty, the situation at the time was described, "Formerly floods were filled in China. People lost the residence so they lived in hills and trees." At that time, there were floods and people were struggling in the waves.

I saw there were ships and silhouettes of people on the peaks of mountains in China. The civilization center was in Kazakhstan, where was the location of the most prosperous period of China. Because this area was near the Kunlun Mountains and the surrounding terrain was relatively high, the great floods were up to more than two thousand meters, flooded the entire Earth, but there were a lot of Chinese people going to the Kunlun Mountains in large floods and survived.

In large floods, the center of the China race and culture was Kazakhstan and the Kunlun Mountains, not the Yellow River basin In old books, we can also find some records, such as: "Wild North Classics," it was recorded that The Jun Emperor went to Tai, in the north of the Kunlun." It was also recorded that"...was thought to be the emperors' Tais, in the north of the Kunlun and the east of Rou Lì.""The Tais of Yao, Ku, Emperor Danzhu, and Tai Er were in the north of Kunlun. Tai was a construction for worship or observation of the planetarium and an important symbol of the ancient civilization. Ancient emperors built Tai in the north of the Kunlun Mountains, showing the Chinese civilization was in the Kunlun Mountains and Kazakhstan at that time and later was gradually migrated to the Yellow River Basin, then the Yangtze River Valley, and other places. Chinese historians limited the scope of activities of the ancient emperors before Yao in the Central Plains, which was wrong.

Kunlun Mountains are the sacred mountain in the continuity of the Chinese civilization. The width of the Kunlun Mountains is up to 2,000 km, ranked in the top of the Pamirs, with the altitude of 5000-7000 meters. The north side is the Tarim Basin at one thousand meters above sea level. The south side is built into the roof of the world with the Himalayas jointly. In legends, the mountain is the home of the Queen Mother of the West.

I saw the south, finding that a lot of Chinese people living in the Himalayan Mountains also survived.

The flood was overwhelming and rare in the world. It destructed white civilization in Europe as well as the Oriental civilization. But because of the Kunlun Mountains, most China ancestors survived in the great flood and thus more oriental culture was retained. Fortunate enough, our ancestors left us with a rich cultural heritage. China's ancient civilization was very profound. Survivors inherited the past Hetu, Luo-Shu, I-Ching, Tai-Chi, Bagua… etc. The ancient culture left parts of them. These cultures handed down made our Chinese nation become an ethnic with a long history and deep connotation.

In fact, more things were left on the ancient. Unfortunately, their successors could not understand. Fewer and fewer were spread in the historical de-velopment and some lost. Now many Westerners know Oriental civilizations of China are very mysterious. There are many handed ancient things even Chinese people themselves do not understand. After the China culture communication to the west, the Oriental way of thinking, is becoming westernized so that it is even more difficult to understand the things the ancestors handed down. Because these ancient civilizations go a completely different road from modern civilization, ancient Chinese civilization cannot be understood in modern way of thinking.

The flood was waning and civilization reappeared on the ground.

Section 21
The Mayan Civilization and the Mongolia Immigrants from the Destructive Great Flood

I saw there were also some people surviving in the Rocky Mountains and the Americas. The people living in the vicinity of South and North America were the yellow race. American Indians are the yellow people. They were sunned upon to be mistaken.

The Americas Mayan civilization was very brilliant, representative of the ancient American civilization. Mayan knowledge of astronomy, calendar, mathematics, agriculture, architecture, culture, and art is very alarming. It has been wondered who created the Mayan civilization?

Lots of people think the Mayan civilization was related to Mexicans; in fact, there was no relationship between them. The Mexican is a hybrid of Spanish and indigenous people. The Mayan civilization was handed down from prehistoric times. Mayan culture is directly related to the Mongols. Before the Great Flood, the yellow people living in the vicinity of Mexico's Yucatan Peninsula was one of the Mongols and they created the Mayan civilization in Central America land.

In the thirteenth century, Genghis Khan and his sons swept across the Eurasian continent by virtue of the iron heel of the vertical and horizontal world to build a huge empire. In the previous history of human civilization before the great flood, Genghis Khan's ancestors created a very brilliant Mayan civilization in Central America. Mongolia reached the pinnacle of military force as well as civilization. It is a really powerful impressive nation.

However, when the big flood came, the Mayan civilization also doomed. The raging flood destroyed the Mayan civilization, including many prehistoric civilizations in the Americas. Americas became a vast expanse of water and people struggled with devastating waves. Most of the prehistoric Mayan were flooded by the big flood and only a small number of Mayan people fled to the nearby mountains and survived.

"One day in the distant ancient, rainstorm dipped, flash flooded, there were oceans on earth, alpine hidden into the water, and humans became fish and shrimp..." This is an ancient legend about the flood spread among Indians in the Americas, Mexico Valley. After flood waters receded, the people who survived returned to hometown. The city was doomed. The survived Mayans became a small number of Indian tribes.

Although they rebuild homes destroyed by floods but failed to reach a height of their ancestors' civilization. They only remembered part of their own understanding of the Mayan civilization but could not recover all of the Mayan civilization.

Because all of the production data were destroyed, people restarted a life of primitive society. Since the environment in which culture existed was destructed, many cultures could not be clarified to be passed. With the Mayans' updates from generation to generation, descendants cannot understand the real essence of the prehistoric

Mayan civilization. The Mayans could only build the post-Maya civilization on the basis of remnants of memories. Even so, the Mayan civilization left is still surprising. An advanced calendar was recorded in classics. It was so accurate that there was the error of only one minute per year, which means there was a day error around 1500 years. This was clever than the Gregorian calendar recognized by the Church in Europe in the beginning of the 16th century.

Like ancient Egyptian civilization, more than once floods were recorded in the Mayan classics. The human history could be traced back to hundreds of thousands of years before the Great Flood. The Bible recorded only a great flood- the great flood in the Noah's Ark era.

Mayans had a profound understanding of the planets in our solar system orbit. Through observation of the orbit of Venus, they calculated the cycle of Venus moving around the Sun was 583.92 days, very close to the calculated results of the contemporary.

The generation of the concept of "zero" was a breakthrough in the ability of an abstract in mathematical operations. The Mayans applied "zero" to mathematical calculations earlier than the Europeans in at least eight centuries, thus made the Europeans greatly surprised.

Then, when Mayan priests told their culture classic knowledge to the visitors from the other side of the ocean, Spanish colonists felt so shocked. Everything Mayans said does not exist in the Bible and even some of the content is also contrary to the Bible. For ex-

ample, the Maya had known that the planets in the solar system turn around the sun and this was the "heliocentric" theory Bruno announced before. Wasn't it departure from the "geocentric" theory?

How could "God's people" allow this kind of "deviant" speech? This frightening remark was simply intolerable for the army bishop! Even European Bruno was burned at the stake for "being deviant," not to mention those Mayans!

Spanish missionaries were scared of the profound insight of the Mayans. Thus, thousands of Mayan culture classics were burned by those religious fanatics.

Mayan history, culture, philosophy, and scientific achievements were all reduced to ashes in a short period of time and only four manuscripts were left. More outrageous is that those "messengers of God" introduced the most shameful stake post in medieval Europe to Americas. Mayan priests were tied to the stake and died in the raging fire.

The Mayan science knowledge was only secretly spread in the priestly class and civilians had no right to learn. With the death of the priestly class, the Mayan civilization was truly lost.

Spanish colonists' evil was equal to Qin Emperor's "Burning of the Books." Religious evil caused by the narrow bigotry of religious fanaticism shows the most disgraceful aspect of Western culture. When the missionaries repeatedly accused the Mayan scriptures were "devil's activities," they did real devil actions.

IV

Starship struck the Earth and Houyi Shot down Nine Suns

Section 22
The Alien Fleet of the Ancient Times Invaded the Earth for Ten Days

Houyi and Chang-Er were the Moon Gods, responsible for the day-to-day maintenance of the moon, to ensure the normal operation of it. After the Flood, Houyi and Chang-Er moved the moon back to the track again. So, the night of the world had light again. I saw that the Chinese people survived gradually migrated from Kunlun Mountains to the east.

Kunlun Mountains were not very suitable for human habitation, so after the great flood the Chinese people migrated to that area of the Xinjiang desert, where was a fertile land then. Modern archaeological research has proven it. The Swede Folk Bergman and B.Bohlen had organized the delegation to scientific investigation in northwest China. They expedited from Suiyuan westward to Xinjiang, finding many sites were located at the edge of the desert or the barren places. Therefore, they supposed the local climate underwent significant changes then: There turned out to be the lush and densely populated lands and was gradually abandoned and became desert today.

When there was multiply population, living space was gradually crowded and some people left Xinjiang and opened up their homes to no-man's

lands of the East. After the lands in the east were flooded, the creatures were doomed and all the livings were dead. The lands were lifeless without any signs of human habitation, becoming no-man's lands. When the population moving eastward increased and there was a shortage of living space, part of the people continued diverting toward the east. Nearly a hundred years, China population continued diverting toward the east, so there were gradually signs of human habitation and footprints in the land of the East.

Later, the Chinese continued eastward. In the period of the king Zhou, there was the story of "Taber Ran to the Wu" and Taber became the ancestor of the Wu culture and founding.

Humans just experienced the catastrophe of the destruction of the world and productivity was back to the primitive society. While ancestors tried to redevelop their home, a bigger trouble came.

One day soon after the flood waters receded, a huge alien fleet suddenly appeared in the solar system. They inquired about the intelligence to took the opportunity to attack and occupy the Earth by taking the advantage of humans suffer from disasters, unable to resist, to obtain the maximum benefit at minimum cost. The alien fleet consisted of ten huge circular UFOs with the diameter to several kilometers, floating in the air like a small city flying on the sky. When they flew over the ground, they could cover a town. The shocking scenes were as demonstrated in the movie "Independence Day." There were a lot of small flying saucers in every major UFO to perform small tasks.

The alien fleet was a multi planet coalition force from multiple extraterrestrial planets, composed of several different aliens. The greatest purpose of their invasion of the solar system was to turn Earth into their common colonial.

The evil aggression was from the ground so that we need to understand the aliens.

Although aliens were also humanoid, the proportion of their limbs and body were not the same with us. The most commonly seen was referred to as the "little gray man," who was the negative alien with the height of about 1 to 1.2 meters and thin bones. They looked very thin and small. Their arms were slender but their legs were short and thick. They had webbed feet and unusual big heads compared to their bodies. Their brain capacity seemed even larger than people on earth. Their heads were bare without hair. Their oddly big and dark eyes looked scary. Their chins and noses were pointed. Their nostrils were two seams. In short, at first glance, they were ugly and horrible.

Although aliens mastered a lot of high-tech than humans, they were in fact low class bio between animals and humans. From a physiological point of view, because we human were made by God, we have the same image with God, thus closer to God. The aliens

were made by God with the image closer to animals. It is because we are created by God, so our body structure is perfect in this universe and the aliens are envious!

Aliens can be divided into many kinds in accordance with the image of humans and each kind is different. However, they do not have special human body structure (human bodies correspond to layers of space in the universe), nor thoughts. Aliens have been found on Earth as many as 67 kinds and their size appearance varies. There are good positive aliens and malicious negative aliens.

In the past, there were no people on Earth. Regardless of updates of Earth in the past, the main life was aliens on Earth.

When their civilization was developed to the late, some of them went to other planets and multiplied, becoming aliens So the aliens came to Earth, just like back home. Of course, some of the aliens were created from outer planets. But this time, God has given us human with the "home."

The alien alliance gained enough intelligence and then decided to send a multinational coalition for invasion of Earth by the advantage of a major disaster on Earth. As Western colonialists invaded the Americas and Africa, the first colonial step was armed capture. Thus, the ancient people saw the suns shining blinding light appear suddenly in the sky, namely a huge flying saucer.

I see the 10 huge UFOs suddenly appear in the sky. Each is as big as a town, flashing angrily and arranged in two rows. They quickly fly in the air. What a shocked scene. Whoever sees it feels extremely strong shocked!

Our ancestors were slash-and-burn than and they even did not know UFO, mistaken ten suns harming the earth. The ten suns were like light balls, sending a powerful heat light to bake earth and torment creatures. Crops were burned and withered. People had no food.

This scenario would also shock even modern people, not to mention ancient people in the Stone Age. Even modern people cannot overwhelm this aggression, not to mention the ancestors in a primitive society. How could they resist the aggression of a highly developed alien? They could only allow the alien UFO light sunburn Kyushu and rage among humans.

Therefore, it is reported in "Huainanzi: Astronomical Institute," in Yao, 10 suns appeared, burning farmlands and vegetation and the people had nothing to eat." This is the myth "10 suns appeared."

When aliens launch a night attack, the bright radiance was more shocking. In the dark night sky, ten glorious and extremely huge UFOs flew low and ancestors thought ten moons hang on the sky. During the day, the "10 suns" poisoned people. At night, the "10 suns" attack. So there was Indian sayings, "ten suns and ten moons."

Mexico is famous UFO hotspot country and there are a lot of UFO sightings. Aliens are particularly interested in Mexico. Therefore, in ancient times, they established their invasion base of Earth in Mexico. Their UFO could

not only fly to the sky but also enter the water. Gulf of California in northwestern Mexico is a narrow edge of the sea in the west side of the North American continent. This bay is long, narrow, and deep into North America. Its ocean and current water is unique in the world, becoming the best base aliens selected.

Gulf of California is also known as the "Tang-Gu" or "Gan-Yuan" in the Classic of Mountains and Seas. Aliens' saucers flew in Tang-Gu. When ancient forefathers saw them, they thought as ten suns in this sea-bathing generated from the more ancient solar observatory official "Siho." Because of the misunderstanding of the Classic of Mountains and Seas: Wild South Classics," Siho was the wife of Emperor Chun who generated ten suns. In the Classic of Mountains and Seas: Wild South Classics," it was recorded that "there was Fusang on Tang-Gu and ten suns bathed in it in northern Hei-Chi. There is large wood in the water, nine suns under branches, and a sun above branches." Tang-Gu is Mexico today, where UFO can be seen in Mexico.

Many people have questions about that living in the water may cause being drowned by the sea water! In fact, the alien's technology has been able to break through some space and their flying saucer is able to break through this space into another space. Here in this space is the sea, but after breaking the space into another space, here is not the ocean, but the place people can live. I sigh and feel less about the existing technological standards of our humanity compared with aliens' high-tech.

In the face of the strong attack of

the aliens, people on the planet still launched a revolt. Such resistance is as if the year-old children resisted a powerful boxer. The Ugly Woman is the heroine of the "anti-Japanese" and we must not ignore her because of her ugly name. She stood up to the alien invasion and ultimately sacrificed in the mountains of the eastern half of North America. In the Classic of Mountains and Seas: Overseas South, it was recorded that "the ten suns killed Ugly Woman after she was born for 10 days in Northern Husband Country. She covered her face with her right hand. The 10 suns live on the sky and Ugly Woman stayed on the mountain." In the Classic of Mountains and Seas: Overseas West, it was recorded that "there was a person in blue with a sleeve covering the face, named the corpse of Ugly Woman."

But our strength is much less than aliens'. Do we humans can only be trampled upon?

While aliens hold high the glasses to hail the victory, they were counterattacked by the people in the ground, our friends of interdependence. If we fall of the terrestrial world, the ground will be the next target of the alien.

Originally it is not a problem to get rid of the people in the ground, but the planet Venus grabbed the land, causing the earth flip and the world under the ground suffer from floods, when the civilization is not as before, they could not fight against the planet Venus. Aliens again dominate and were ready to attack the terrestrial world.

The people on Earth are facing to be subjugated.

Who we can rely on at this time?

Who is going to help us?

Wish the heavens bless us!

At this point, a Chinese god hero is coming soon.

He will save the people being in the firelight.

Section 23
Houyi Shot down Nine Suns and Attack the Alien Fleet

In Tang Valley there was no wave.

In another space, there was the command flagship UFO of alien alliance where alien chiefs were discussing offensive battle plans. The alien commander-in-chief said with the alien language, "There is no need to consider the people on the ground. Now we have to consider how to attack the underground world, to gain maximum benefit with minimum cost!"

Suddenly, a white god flew into the venue and then stopped at a round table in front of midair. Aliens surprised and then pulled a "gun" at the man god, asked, "Who are you, and how did you enter and come to us?" The white god said, "I'm Luna Yi. If you bully does not stop, you will die by the curse!"

The alien commander-in-chief realized what he meant and said with alien language, "We are God, not afraid of you and you curse!" Then, he shot a laser beam from the gun to Yi's chest. Other aliens also shot at the same time. The deadly beam fired together. If Yi is the ordinary man, he had already become a focal corpse. But Yi's stature suddenly faded. Clutter laser beam has made the walls of the conference room full of a dozen big holes.

Aliens were to be shot again but there was no human here. This made them very surprised but the smooth military situation made them forget about this incident and planed their aggression blueprint.

Arrogant aliens mastered some high-tech and thought themselves as God. They looked down on the gods. But God grasp the real high-tech much better than aliens. The outcome can be imagined if aliens want to fight against gods.

The war into the world under ground began.

The entrance into the ground under the world is hidden in the channel at the north and south poles. The South Pole is colder than the North Pole and the wind of the South Pole may kill people. The South Pole is known as "the hometown of the blizzard." The cold ice sheet of the Antarctic is like a machine manufacturing cold wind, all the time cooling air and nurturing storms with snow. When the terrible polar storm rages, the wind is huge and darkness is everywhere. Snow and ice carried sands and rolled down from the slippery ice slope, just like the galloping torrent.

UFOs in this storm also have crash danger like leaves in the rapidly flowing water. Therefore, aliens took the strategy of surrounding instead of attacking in the Antarctic entrance and sent three UFOs waiting in the coast of Antarctica. Once the vehicle escaped under the ground, aliens would surround and annihilate it with the triangle horn form. Arctic entrance is more suitable as a breakthrough compared to the South Pole. So, aliens sent six flying saucers from the Arctic channel to attack the world under ground. Flagship UFOs stayed at the base, directing operations.

Aliens planned well but they did not know that the mantis stalks the cicada, unaware of the oriole behind.

In the early morning, the sun is not up yet and nine alien flying saucers set off. Three saucers moved to the Antarctic and six to the Arctic. They wanted to attack the people under ground intensely when they slept.

In this case, Yi (Houyi) and Chang-Er realized the attack plan of aliens. The people under ground had no ability to destroy the invaders, so they had to start immediately the injured moon, slowly left the track and flew first to the Antarctic.

Yi and Chang-Er cooperated. Chang-Er was responsible for driving the moon and Yi was responsible for offensive and defensive. Yi was worthy of being a great marksman. He aimed the aliens' UFOs of low altitude in the high altitude, targeting and attacking. A giant of the light jetted out from the moon and a UFO floating in the Antarctic off the coast suddenly exploded and instantly enveloped with the bright light gushing out. The UFO burst to pieces in the sky and scattered, looking spectacular.

The aliens in another two UFOs were in shocked and confused, while Yi shot two giant lights and hit them. The two huge UFOs quickly became two groups of giant fire, emitting a dazzling brilliance in the dark and crashed in the Antarctic's seas.

They attacked for the weakness of the enemy to cut off the backing of their enemy. Then they assaulted fortified positions for victory. Yi let Chang-Er drive the moon rapidly to the Arctic. At this point, the North Pole had fires and the fighting sound came. The war

of aliens and the people under the ground began. Laser meteors were all shot to the entrance of the Arctic. The people in the ground guarded the entrance with powerful firepower as guardian of the thoroughfares. If one man guarded the pass, ten thousand could not get through. The aliens difficultly invaded.

The fighting was so intense that the aliens did not notice there was a bright moon on the sky. A giant light shot in and a UFO burst in the enormous flame and scattered, slammed into the Arctic Ocean. Other aliens on the UFO heard this loud noise and felt shocked with the situation, and then they noticed there an additional moon in the sky.

Yi shot again and again and every hit would be in. Thus, several huge UFOs turned into a fire in the sky as stars flying, falling in the iced sea.

The aliens were much panic. They did not expect opponents attacked behind them, while several remaining stampers hastily turned their guns to hit back at the moon. But the moon was so high that it cracked down in a sharp tendency. The huge light energy attacked the aliens and that the UFOs became splendid fireworks in the night sky in the Arctic. Clouds of flaming crashed in the Arctic ice and snow. Five stampers were completely annihilated. Only one huge flying saucer and several small flying saucers escaped desperately, they were shaking and flew desperately to the North American base. Yi let Chang-Er drive the moon and flew further into the alien base direction.

At this point, the sun rose and the aliens had sent defeat news to their base. The aliens' communication equipment had long been beyond the imagination of the people on earth. They used thinking sensing, which was like the fax machine on Earth and could send out text or images by thinking without any condition limits. Aliens' ground receiving station also received information with the thinking sensing function. As soon as the alien commander-in-chief got the news of the defeat, he was stunned. He never thought about the result. He thought how to do. Soon he ordered to attack and rescue his army.

Flagship UFOs broke and leaped off the sea almost without sound sea straight to the north. The warm sun shined into the cabin but the aggressor did not feel it's warm but felt a sense of grief and indignation. Soon, it saw a stamper wavering and flying back. It flew with a little voice but obviously it seemed to be very afraid. At the same time, the moon on the sky moved exceptionally quickly, getting closer and closer.

"How can I let it prevail?" Yi thought. The god light shot with radiance. The UFO burst and burned like a huge fireball in the sky with reflection of the sky. It slammed down from the sky, shining the light of the flame which was abnormally gorgeous. Immediately, it became a heap of chaos fire, plunging to the ground. I the crashing loud sound, it fell in the land of the west side of North America.

After the people on the ground saw the giant fire falling from the sky, they ran to see, finding that it was very huge metal three-legged black object debris and thinking it was the sun. So the ancients called the sun as the "Golden Bird".

Since then, "Golden Bird" became sun's alias. In "Huainanzi: Spirit Section," it was said that "there was a crow in the sun." In Guo-Pu annotations, it was said that "there was a three-legged bird like a crow, which might be the driver of the sun." Birds have always been two-legged. Why there was a three-legged bird and why was this bird still in the sun? In fact, the huge UFO was stamper and there were a lot of small flying saucers showing crow black. When they stayed on the ground, they used three stands. As a result, they were known as the "three-legged bird" or the "three-legged crow." Also there was a legend that the sun was the three-legged crow because the large UFO was also like a three-legged crow.

The flagship UFOs saw that ship stamper exploding and nine stampers became dust. The commander-in-chief felt terrified and knew that this battle would end. He thought to run away to save their lives. In the occasion of shock, it ordered the UFO to turn around and simply escaped towards the deep space.

When Yi saw the enemy UFOs escaping, he asked Chang-Er to catch up them. However, the huge moon's speed was much slower than UFOs. Seeing the enemy flying saucer escape, Chang-Er hastily drove the moon to catch up them. But desperately, she opened the key of the moon, namely the Excalibur, and broke it. When Yi and Chang-Er saw the artifact broken, they felt greatly distressed. At the same time, they saw the enemy UFO escape from the firepower of the moon but they were helpless. They could only sigh for the providence nature!

Yi wiped out nine UFOs and one escaped. Thus, in future generations, there was a legend of "Houyi shot sons and nine out of ten were shot." In fact, Yi and Houyi were two different people, but they were equally good at shooting; therefore, descendants mixed up their stories. It was Yi shooting, not Houyi.

Then, Yi and Chang-Er tried to direct the moon back to the right track, inserted the broken sword into a huge stone in the underground world, and left the prophecy: "The person who can pull out the sword will be the one to repair the Excalibur and the moon.

They were also dropped into the mortal world for dereliction of duty. Because the gods in the universe thought this was too sanguinary to slash most of their divine power, they could not live in the moon anymore and so they came to the world.

In the world, Yi first assisted a prince of Emperor Yao and helped the prince to set off to the war with the rest divine power; he killed some of the fairies harm to humans. The prince was therefore greatly credited, appreciated by the father (Yao II), and eventually inherited, becoming the third Emperor Yao. So Yi served as the third Minister of the Emperor Yao. Yi earned meritorious for auxiliary of Yao III's throning, Yao III designated Yi as the monarch of a vassal state then.

Therefore, it is reported in "Huainanzi: Astronomical Institute," "human-face, horse-foot monster, long-teeth monster, nine-head monster, phoenix, large wild boar, and python were harmful livings to people. Yao commanded Yi to kill long-teeth monster in the domain, kill nine-head monster over the fierce water, hunted phoenix in Creek Qiu-Qing, shoot ten suns, kill human-face, horse-foot monster, cut off python in Dongting, and capture large wild boar in Samrin. People were jubilated and regarded Yao as their emperor." After Yi and Chang-Er's death, the king's position was inherited by their children and grandchildren.

Section 24
"Classic of Mountains and Seas," the Geographical Encyclopedia Four Thousand Years Ago

"Yu Gong" was written for Emperor Yu Tames the food. Dayu and Boyi wrote Classic of Mountains and Seas to record the lands, seas, scenery, and cultural relics they experienced for thirteen years in water management for transmission to future generations. However, as described in the Classic of Mountains and Seas, there were many bizarre and sensational plots so that people could not understand it and it was dismissed as an absurd works. Therefore, the "Classic of Mountains and Seas" became a history and geography book with a wealth to the myths and legends.

The "Classic of Mountains and Seas" was a wide-ranging world masterpiece with deep cultural deposition. It can be said, "Classic of Mountains and Seas" could be regarded as the world's cultural treasure and it was the perfect valuable information to study the history and geography of ancient times.

In contents, the "Classic of Mountains and Seas" contained history, astronomy, geography, ethnicity, folklore, myth, religion, witchcraft, animal, plant, water, property, mining, traditional Chinese medicine, and many other aspects, being content-rich, all-encompassing to be called as the China's classical works with the richest contents.

In time, it spanned three human civilization periods and kept our national remnants of memories about history, geography, and knowledge from ancient times to the Shang and Zhou dynasties as well as Chinese civilization and the origin and development of culture. When Yu and Boyi wrote "Classic of Mountains and Seas," they referred and integrated many previous history, geography, and prehistoric civilizations into what they had seen and heard at their times. There were much ancient information that they themselves could not realize, not to mention for the future generations.

On space, it described almost the world's countries and geographic information in the period of Dayu, as well as the natural ecological environment in the survival and development of world. This is associated with the rule range of the yellow race. The Kunlun Mountains, the Central Plains, and the Pacific Rim (including Southwest, Northeast, North and South America) were the countries of Chinese ruling.

Dayu traveled to these places so he understood some information beyond and recorded in the "Classic of Mountains and Seas." Therefore, the geographical scope of the "Classic of Mountains and Seas" is Kunlun Mountains, the Central Plains, the Pacific Rim, and the more distant exotic areas.

In the Western Han, Liu-Xiu recalibrated the book and set up eighteen chapters from the original thirty-two chapters. He wrote in "Shang Shan Hai Jing Biao" that, "when Dayu controlled the flood, he also drove away animals, named mountains, classified vegetations, and distinguished water and soil with four mountains to surround the area to deserted rare places. In 'Shan-Hai-Jing,' the nations about one hundred miles away, five hundred fifty mountains, three hundred waterways, and the geographic relation among the landscapes of these nations, customs, folklores, and important properties were described. It also included hundreds of historical figures as well as activities or lineages of these figures. Among them, there were abundant materials of original myths and primitive religions. A lot of diverse Chinese ancient legends could be preserved thanks to this book.

In the book, it was stated that Yu and Boyi had been to every corner of the world when controlling the flood so they met a lot of special people. They had been to foreign countries but modern people do not believe that and they think it as the rich imagination of the ancients.

In past dynasties, the awareness of the geographical value of the Classic of Mountains and Seas had experienced a zigzag process. In Eastern Han Dynasty, the famous flood control expert Wang Jing controlled the river and one of the reference books the Emperor Ming gave him was the "Classic of Mountains and Seas." When Li Tao Yuan wrote "Shui Jing Zhu," he cited from the "Classic of Mountains and Seas" at more than 80 items. Then, "Suishu Jingjizhi," "JiuTangShu Jingjizhi," "Xīntangshu Jingjizhi," and Wang Yao Chen's "Chongwenzongmu," the "Classic of Mountains and Seas" was included in geography class book of the history. Ming and Qing Dynasties were the period when the value of "Classic of Mountains and Seas" was depreciated. It was said to be "too many supernatural things" and "it was difficult to test mountains and rivers here."

To modern times, Gu Jiegang wrote"Wu Zang Shan Jing Shì Tan" and published many extremely insights, allowing people re-aware of the scientific value of the" Classic of Mountains and Seas." Subsequently, Tan Qixiang wrote "ShanJing Heshui Xiayou Ji Qi Zhiliu Kao," affluxing tributaries of the river downstream in therich river data of "Shan-Hai-Jing" into the "Beishan Jing" and paralleling it to research out the most ancient old course of Yellow River.

The publication of the article further established the scientific status on the geography of "Shan Hai Jing" and "Wu Zang Shan Jing." The part with the most geographic value in "Classic of Mountains and Seas" was "Wu Zang Shan Jing," in which the content was elegant and formal in describing mountains and rivers; although mixed with myth, the proportion was small so that it was no doubt an early geography book. As for the historical value of the "Classic of Mountains and Seas," because it described the things before true history times fuzzily, descendants were difficultly to research the authenticity. Thus, it can only be a myth or a legend. But Classic of Mountains and Seas reserved a lot of history and geography information in the prehistoric Chinese nation in the form of legend, becoming the supplement for the absence of the ancient parts in the Chinese official history.

In the spreading process of "Classic of Mountains and Seas," there is scattered, lost information as well as addition and modifications several times. Therefore, there were more than one author in "Classic of Mountains and Seas" now we see and it was not written in only a period of time. Those long times distant memories were passed down from generation to generation and it was bound to be fuzzy and mixed. Therefore, there were some errors in the "Classic of Mountains and Seas" and descendants need to carefully verify its contents.

Section 25
Chang-Er Flying to the Moon-she was not an ungrateful person and flew to the moon in a critical

Dayu was elected as the heir of the Emperor Shun for his successful flood control and inherited the throne after the death of the Emperor Shun. Yu did not pass the throne to his son but he still followed the abdication system to elect Gaoyao as his heir. However, unfortunately, Gaoyao died before inheritance so Dayu elected Boyi. Although Boyi was Yu's legal successor, he failed to inherit the throne after Yu's death. Three years later, the son of Yu took away the throne.

Qijifu became the Emperor and seized Boyi's legitimate right of inheritance. It was opposed by other tribes. The tribe Youhushi refused to accept him. Youhushi thought Yao, Shun, and Yu took democratic abdication so they could not destroy the tradition. In order to maintain the monarchy, Qijifu led a military to crusade against Youhushi in the place Gan. Before the combat, Qijifu held the oath-taking and said, "six army soldiers, Youhushi insulted the gods, acted in guards, and did not take the human right; therefore, God will destroy Youhushi. I set off to fight them out of complying with God's punishment. Now I order the left army to attack their left, the right army to attack their right, the middle army for frontal assault. Ministries and soldiers must obey orders and move forward. The brave will be rewarded and the disobedience will be killed, with guilt to their children!" Because six Army soldiers heard destroying

Youhushi was God willing, achievers were rewarded, and the retreat people would be killed, they courageously fought. The small Youhushi could not stop Qijifu's powerful offensive. After a bloody battle, the entire tribe was defeated. After Youhushi was eliminated, the tribes of the Yi Xia were surrender with Qijifu. Qijifu further consolidated the throne. Then, China had the political system of throne hereditary and the world belonging to China.

After Qijifu conquered ministries, he convened princes in JUNTAI. The assembly was unprecedentedly grand and Quartet princes came. The tributes around the country were filled over JUNTAI. There were huge dancing bands, meat, and wine. It was wealthy and extravagant. After the assembly in JUNTAI, Qijifu patrolled, ate, drank, and had fun. He was not totally like his father Dayu. He became more and more dissolute. Then, he did not notice and care for public affairs but merely enjoyed pleasure. Internal strife of his five sons occurs for the throne inheritance rights. Royal power weakened.

After Qijifu's death, his son Taikang succeeded to the throne. Affected by his father, Taikang ignored political affairs and indulged in hunting. All industries were waste and discontent was everywhere. This provoked a revolt emotion of various tribes of Yi Xia. Thus, a fierce remonstration began. Houyi, the leader of Dongyi's Youqiongshì revolted, expelling unpopular Taikang. Emperor Yao had ordered Yi as the monarch of a vassal state and the state was Dongyi's Youqiongshì. After Yi and Chang-Er's death, the king's position was inherited by their children and grandchildren.

Two hundred years later, in the Xia Dynasty, Yi was reborn into this vassal state as the monarch and his wife was also Chang-er. Because he was good at shooting, he was called as Houyi (Yi was also a marksman). Since Taikang was corrupt, soldiers and civilians were rage. Houyi sent troops to cut off his return while Xia carried his family for hunting. Then, Yi left the troops to expel him and refused Taikang to return to the capital. Houyi crowned Taikang's brother Zhongkang as emperor and Houyi served as the prime minister for assisting the affairs of state. Houyi was not opposed for that, which reflected Taikang was not supported by the people then.

Because of the karma, Houyi couples could be together again. Houyi handed over the prime minister post to his crony, an ambitious man called Hanzhuo, and he went to the Kunlun Mountains with Chang-Er for accomplishment of the monastic life. Hanzhuo was the bad son of Han Country's monarch Boming. He was shut out by his father for his contemptible and foxy characteristic. Houyi gave a shelter for him. Hanzhuo pretended he was faithful to Houyi and received Houyi's trust so that Houyi appointed him as the prime minister of the country and assigned Hanzhuo all Government affairs for processing.

After a decade, Xiwangmu, the elder God in Kunlun Mountains told them that they could leave the mountain and gave Chang-Er two magic pills for emergency use.

In the decade they departed, Hanzhuo managed his personal forces and bribed to hollow out Xiawang. Houyi's return made Hanzhuo very afraid so he staged a coup and killed Houyi, Xiawang, and their families.

The heinous Hanzhuo led his troop to attack the Yi House for massacre.

Hanzhuo had come prepared and launched a surprise attack. Houyi unprepared and caught off guard. Finally Hanzhuo killed Houyi. Logically speaking, cultivators should not encounter this mass annihilation. However, God seemed to want Houyi to have the retribution his shooting and killing suns in the previous years so that he must encounter such attack and die. Houyi suffered calamities for humans!

Hanzhuo forced Chang-Er and Houyi's family obeyed him. Chang-Er did not want that and shouted with rage against him. Hanzhuo was angry so he ordered his subordinate killed Chang-Er.

Chang-Er escaped and ate two life elixirs in emergency. After eating, she completely changed, her body was lighter, and she flew up involuntarily.

When Hanzhuo and his subordinates hunted Chang-Er, she was left the ground and floated to the moon. Finally, she entered the moon. When Hanzhuo's subordinates saw Chang-Er fly away, they felt scared and afraid on the spot.

So, this world had the story of "Chang-Er Flying to the Moon."

Chang-Er flew to the moon and the ancients could describe the matter by "Chang-Er Flying to the Moon." If, as the legend, she stole the elixir, she did not need to flee away. This also conveyed the truth for us.

In "Classic of Mountains and Seas," it was written that Chang-Er was the ungrateful person because she betrayed Houyi to flee to the moon. She stayed lonely in the moon everyday, looking at the Earth and crying. After a long time, she spent her all day in tears and could not suppress sadness. So, she wrote a poem:

Battle caused cold mist, being with unbearable grief in terrible and bloody life. Flying to the moon in the blue sea and sky, people referred Galaxia to be Yu Chan. "Reading Chang-Er Flying to the Moon," RainDu

The frenzied Hanzhuo cooked Houyi, forcing Houyi's son to eat it. Houyi's son could not bear to eat so he tragically suicided. So Hanzhuo officially replaced Zhongkang as the emperor and began his despised usurpation ruling. In later official history, his name was almost not mentioned but referred to be a "prime minister" for the ruler then because he was regarded to be a usurper.

Chang-Er returned to the moon while Yi was still in the human reincarnation.

Section 26
Shaukang Recovered the Nation and Houyi Reincarnated on the Earth-endless Resentment

In the bloody massacre, a wife of a prince survived from the disaster because she was pregnant and stayed in her parents Yourengshi tribe. Later, she gave birth to a boy, named Shaukang, who was Houyi of reincarnation.

After growing up, Shaukang became the pastoralist in his uncle's home tribe, responsible for the cattle and sheep husbandry. Hanzhuo knew his whereabouts and sent people to kill him. Yourengshi was unable to protect Shaukang so he was forced to leave to the Youyushi tribe secretly alone.

Youyushi was Yu Shun's head tribe, which had been respected by the world since Yu inherited Shun. At the time, the leader of Youyushi was the descendant of Yu, Yusi, who had been dissatisfied with Hanzhuo's tyranny. He felt Shaukang held extraordinary tolerance like Xia ancestors, so he enthusiastically took him with warm hospitality. Yusi assigned Shaukang to be the cook of the tribe and taught him how to manage the property and the ability to lead the troops to fight.

Soon, Shaukang became a civil and military man. Later, Yusi felt Shaukang was faithful and had capability, so he let him marry his daughter Ertao and gave him the fields of ten miles Lunde and five hundred dollars to support his development of forces. Since then, Shaukang gained a foothold. Based on the tiny piece of the site, he actively ac-cumulated strength and made the pre-paratory work for national recovery.

Under Shaukang's governance, Lunde's production was developed and society was stable. Also, here was flourishing population, becoming the well-known "Leguo." The Xia people were scattered in the mountains. When they heard Shaukang was the offspring of Xia and Lunde was populous and prosperous in his governance, they gradually came to this place. With the support of the people, the power of the Shaukang had been growing up.

When Hanzhuo killed Houyi, the veteran of the Xia Dynasty Bomi did not follow Hanzhuo but fled to the Yougeshi near Dezhou, Shandong. He relied on Yougeshi's power to win the adherents of the Xia Dynasty, Zhenguanshi and Zhenxunshi, over by any means and organize strength to prepare for the fighting of future generations ruling of Xia. At this point, Shaukang had accumulated strength in Youyushi, ready for the recovery of the nation. Bomi connected with Shaukang and led his men to be the followers of Shaukang, forming an army of nation recovery. Shaukang's conditions of national recovery were gradually mature. He took Youyushi as his backing, and then with the assistance of the old veteran of Xia, he differentiated, strove for the people of Hanzhuo, and launched armed attacks to Hanzhuo.

Shaukang's recovery army first captured the vassal state of Hanzhuo's eldest son Hanjiao, Guocheng, and killed Hanjiao. Then he attacked the vassal state of Hanzhuo's second son Hanyi, Geyi, and killed Hanyi after recovery. They captured two vassal states of Hanjiao, which was equal to cut down Hanzhuo's wings and recover the most parts of the Central Plains.

Then, Shaukang attacked Hanzhuo's old base Chateau. Shaukang's army was suddenly close to the town and the battle with strength and impetus began.

Hanzhuo was an old man, unable to go on an expedition for killing. And he had been enjoyed life and ignored the public affairs, people were all complaint about him, so he had to hide in the palace. Shaukang was majestic with high morale. When the Hanzhuo Army and the Xia Army fought hand to hand, the Hanzhuo Army was defeated. The Xia Army surrounded the siege and Hanzhuo failed. Hanzhuo's subordinates saw the hopeless situation and suddenly rebelled Hanzhuo in order to give their families a way. They reached the palace and tied Hanzhuo, sending him to Shaukang. Hanzhuo's regime collapsed. In people's cheering, Shaukang recovered the ancient capital of Xia.

Shaukang stated Hanzhuo's guilt and ordered for capital punishment; his flesh was cut to pieces for lynching. Also, Shaukang commanded to annihilate Hanzhuo's family. The naturally cunning, sinister and ruthless schemer finally got the fate he deserved.

The Hanzhuo regime was annihilated by Shaukang. Shaukang finally recaptured the country ancestors lost. He paid homage to ancestors, appeased the people, rectified the ancient capital, and rebuilt the country to restore the rule of the Xia Dynasty.

From Taikang's failue to Shaukang's recovery, the Si's throne of the Xia Dynasty now finally restored since the bereavement for decades. It was known as "Shaukang's Resurgence." For decades of kingship turpitude, profound lessons punished the rulers of the Xia Dynasty. The aristocrats of the Xia Dynasty learned from their mistakes and no longer dared to be extravagant. After Shaukang's Resurgence, the Xia made efforts in good governance and gradually migrated to the middle reaches of the Yellow River, experiencing a prosperous period for about more than a century.

However, the hatred of Houyi and Hanzhuo was made since then and it was difficult to resolve it. In the later reincarnation, Houyi and Hanzhuo were mortal enemies until today. Strangely they were very much like each other.

In bitter metempsychosis for thousands of autumns, when would both good and bad stop? If hate sources were not painfully stopped, how anger and worry couldn't pass down?

Section 27
Concubine Yang, the Originally Moon Palace People, Recalls the Past with a Song "Plumage"

Chang-Er's level was not high. Every several years, she would be reborn to this world for re-cultivation to return to the moon. She was once Concubine Yang.

Chang-Er was down to earth in reincarnation as Yang Yuhuan. When she grew up, she was a tall and plump beauty. In accordance with the algorithm now, she was one hundred and seventy centimeters high, the absolute height of the model. She was much taller than ordinary maids. In Tang dynasty, plump shape was preferred. And Yang Yuhuan was plump, tall, and elegant. So, Tangminghuang was attracted by her even he was old then.

The Emperor Shun in ancient times became Tangminghuang (Emperor Lilongji) in the Tang dynasty after the reincarnation. Throughout Tangminghuang's life, in the pre-political period, it was a golden age with unprecedented economic, cultural, and artistic prosperity. All countries sent messengers to come for studying their techniques and skills. It was the brilliant "Kaiyuan prosperity." In Tangminghuang's later years, he married his daughter-in-law Yang Yuhuan, far away from morality. The Tang Dynasty declined since then.

In 745, the 28-year-old Yang Yuhuan was canonized as a concubine. She was favored and Tangminghuang did not host morning sessions since then. Since Yang Yuhuan loved lychees, Li-longji ordered to send lychees from Guangdong to Chang'an and many horses died for the long and hurry run. In Dumu's quatrains "Qing Hua Ging Gong," it was created that: Looking back the embroidery city from Chang'an, gates were open from the hilltop. A concubine smiled in the world of mortals, no one knew it was for lychees.

It was recorded in history that "Yang Yuhuan was plump and beautiful. She was good at singing, dancing, and playing the instruments." In order to please Yang Yuhuan, Tangminghuang respectfully invited the famous musician Maxianqi to the imperial palace for teaching Yang Yuhuan playing the instrument. Yang Yuhuan was a smart woman so she was proficient soon.

One day, she took an uncompleted music score which was created by Li-longJi previously to Maxianqi and said, "Could you finish it?" Maxianqi saw it and highly praised the emperor's talent. He did not expect Maxianqi had such a deep knowledge of music.

Why Tangminghuang dreamed he went to the moon and heard "Colorful Plumage"? It was because Chang-Er had been reincarnated but no one knew it. Maxianqi said to Concubine Yang, "Royal empress, this is an exceptional dance music so I could not complete it by myself. I need to select some experts for discussion. At the same time, we must find the person who can dance to express the essence of the song better."

Concubine Yang said, "I have been practicing dance since childhood and there are some people in the palace good at dance."

Maxianqi said, "This is a score adapted from the great music of Brahman in Tianzhu. To continue to write it, it is needed to know the country's culture and customs. Zhangyehu (a national pipa musician) and I failed to grasp the essence of it."

Yang Yuhuan was disappointed and said, "what is a regrettable thing!"

Maxianqi smiled and said, "Not necessarily. Zhangyehu and I have two friends; one is a foreign musician in the Arab region and the other is a musician of western Kangjuguo. With their help, it may not be difficult to complete the composition."

Concubine Yang listened to him and said happily, "Please contact them immediately. I will personally make the dance shirt. After the song is completed, please select a group of folk people good at dance for practice."

The song was quickly made, called "Ziyunhui."

To practice the dance of "Ziyunhui," Tangminghuang ordered eunuch Gaolishi to select 28 women good at dancing, divided into two teams. There was a 13-year-old pretty girl Xieaman with very elegant and soft dance skill.

The Xinfeng girl Xieaman was selected to the palace. She studied at musical house and had primary skills. She could tightrope and basic skills of dance. So she could learn new dance quickly.

Concubine Yang personally summoned her and ordered Maxianqi and Zhangyehu taught her specially. In just a few years, Xieaman's dancing was the best in the palace.

After dozens of changes in "Ziyunhui," it was officially renamed as "Colorful Plumage" and became a famous tune eulogized for future generations. Xieaman became an irreplaceable performer of this sing.

Libai was talented and often invited to the palace for writing poetry. Concubine Yang admired Libai but they didn't have ambiguous emotion. Every time when Yang Yuhuan met Libai, there were two maids of honor following. So the scandal surrounding was just a rumor.

Libai was known as Zhexian and his birth was very legendary. Libai's uncle Liyangbing described the birth of Libai in "Caotangjixu" that "Before Libai's mother gave birth to him, she dreamed Taibaixingjun so she named him Taibai. Fanchuanzheng wrote an epitaph for Libai "Tang Zuoshiyi Hanlin Xueshi Ligongxin Mubei Bìngxu," in which it was recorded that "Libai's family was surnamed as Li. But in Western Regions, they didn't use it and reuse it after Shu."

Libai's parents regarded him as the reincarnation of Tai-Bai Venus, so they named him Bai and the courtesy name as Taibai. Tai-Bai Venus and Chang-Er were friends since they were in the heaven.

In Tempo 15, Anlushan rebelled. Tangminghuang and Concubine Yang fled away without the knowing of all the ministers. On next day, ministers found that the emperor disappeared and there was a chaos in the palace. When the emperor fled, the Prince instigated the mutiny to seize power. The Prime Minister Yangguozhong was killed for the offense of betrayal. It was rumored that Concubine Yang was hanged in Maweipo.

Maxianqi fled in the crowd. On hearing the news of Concubine Yang's death, he was very sad. Suddenly he heard a familiar voice shouting his name from Xieaman. "Why do you here?" she asked. Maxianqi described his fleeing.

Xieaman moved close to him and whispered in his ear, "Let's go to a place with no one. I have something to tell you."

Maxianqi followed Xieaman to a small courtyard. Xieaman closed the door and suddenly knelt down, saying, "Please help us! Chaise Goddess was not dead but in a very dangerous situation. If she was found, she might be dead."

Maxianqi was very shocked. Then Concubine Yang came from the room and nodded to him. She could not talk because her throat was reined and her sound was affected in hanging.

Xieaman told Maxianqi, "We know that the wife of the prime minister stays in Dongying messenger's posthouse. If we can contact her, we could send Chaise Goddess to Dongying for refuge. Staying here may be threatening to her."

Maxianqi promised them and then sneaked into Dongying messenger's posthouse, contacting Yangguozhong's wife and sent Chaise Goddess to Dongying's merchant ship secretly. Xieaman accompanied Chaise Goddess to leave.

Maxianqi saw the vessel far away on the sea, only wishing them to be safe.

In "Old Tang: Concubine Yang Biography," it was recorded that Xuanzong used to order people to bury Chaise Goddess secretly but he could not find her bones. In the tomb of Yang Yuhuan in Maweipo, only a sachet was found so that there was a rumor in the imperial court.

As a result, Xuanzong ordered alchemists to search by ship on the sea to Yamaguchi Prefecture, Japan. They met Yang Yuhuan and gave her two Buddha statues from Xuanzong. Yang Yuhuan unplugged her jade hairpin to the alchemists for gratitude but she refused to return to China.

Concubine Yang stayed in Japan forever, leaving a lot of relics and artifacts. Famous Japanese movie star Momoe Yamaguchi told reporters in 2002, claiming to be the descendants of Yang Yuhuan.

V
Extraterrestrial Civilization-the Human Crisis

Section 28
In the Battle to Defend the Earth, the Two Extraterrestrial Forces in the Galaxy Came at the Same Time

I see the rise of human civilization. In the mountains in Peru and Chile, South America, the culture had been gradually prospered. Then there was the rise of Egyptian civilization in the Middle East. Shortly I see Chinese civilization which became the highest then, and the prosperity of Indian culture.

In the outer galaxy, aliens have been constantly developed and variegated in the long years. They fought and competed for the site. Greed and lust led to star wars. In the Southern Song Dynasty, the aliens in the galaxy are divided into two major forces after years of campaign. They came to the solar system at almost the same time, focusing on the Earth.

The alien's invasion occurred again and the battle of the people under the ground and the aliens began at the edge of the solar system.

Initially, a lot of gods and cultivators in the mountains helped the people under the ground, making the people under the ground won more than the aliens.

But then, the senior God in the universe thought that the earth was gradually corrupted and this period of human civilization was not too far away from the final destruction. To be destroyed anyway, let those aliens bring disaster on earth! If aliens had gone too far, finally they would even destruct the aliens. This was an act of God and Buddhist and Taoist gods and cultivators had to conform to God willing because the Guards and the line would be subjected to punishing.

When the Zhou Dynasty overthrew the Shang Dynasty and being in the Wu King Attacking, many cultivators supported King Zhou at beginning because the early Dixin (King Zhou) had excellent comprehension and great strength and he often broke new ground.

In "Xunzi: Fei Xiang Pian", it was recorded that Dixin was tall and handsome with outstanding capability. Also, he was very powerful and no one could defeat him. In "Records of the Historian: Yinbenji," it was recorded that Emperor Zhou was intelligent, sensitive, and vigorous." The atheling is called as the king. Since Dixin conquered many kings of small countries, he was called as the great king. In fact, Dixin's source of knowledge of God was a very positive life. He was very tall and very wise. It was his willing to govern the nation well and benefit the people. Therefore, many positive lives helped him initially.

But the Gods of the old universe changed the prehistoric arrangements for their need of their historic arrangements, they confused King Zhou with the nine-tailed fox, which was Daji. When Daji entered the palace, the nine-tailed fox was possessed on her. She used ecstasy to control King Zhou and did many bad things in the use of King Zhou. In fact, those things were not out of the willing of King Zhou. The nine-tailed fox would originally like to become immortal, but her mind was used by the Gods of the old universe to destroy King Zhou so pratyaya was created.

After that, the knowledge god and the primordial spirit separated. The knowledge god did not do anything bad in the state of understanding. The primordial spirit was arranged to practice by the Gods of the old universe and he manipulated words and deeds of King Zhou with the nine-tailed fox.

In accordance with the original revere fengshenbang, Jiangziya crown the primordial spirit of King Zhou as God. King Zhou's knowledge God did not do bad things nor practicing but continue reincarnation among people. All the bad things in history were from the nine-tailed fox's manipulation so sin is all owned by her.

Afterwards, she was in reincarnation among people to return karmic debt. Since she did too many bad things, the Gods of the old universe did not allow her to be in dhamma. She wanted to practice in every life but the Gods of the old universe did not allow her. Her last life reincarnation was reincarnation in Hong Kong. She spent her youth in the fashion design industry but died in a car accident.

As a result, to be not good is quite dangerous. So, a lot of cultivators supporting the people under the ground left. Many false gods and cultivators joined the aliens. The people under the ground began defeated. In the Ming Dynasty, the people under the ground had retreated to Earth. Many fights had taken place in the oceans and inaccessible places, occasionally bringing disaster to mankind. The battle also happened in a lot of other spaces and we humans could not see it. In the senior parts of the universe, the positive gods were defeated by the Cthulhu Asura many times. There were more and more aliens in the solar system. The people under the ground retreated to the underground. Later, the people under the ground were completely closed in the ground by aliens and the earth surface was completely occupied by aliens. However, the wars aliens offended the people under the ground failed until today. It was because the executive gods only wanted aliens to bring disaster to the world, instead of the ground within the world. Aliens were also controlled by the cultivators conducting the order of Gods of the old universe. (Taiwan and China are in such an arrangement, with the blood in Taiwan.)

Section 29
The Mystery that the Aliens Travel through Time and Space-the truth of another space

The universe was so vast and aliens' planet was so far away from us. By the existing technology of our humanity, even if moving at the speed of light, it would take many years to arrive it. Sometimes it even took decades. How aliens solve this problem? It was because the UFO and space-time.

The speed of aliens' UFO is amazing. Why is it so? It was because aliens had exceeded some additional spaces so that their UFOs could easily break through this space into another. In a fast time and space, the UFO ran very far for a while. Besides, on time and space distance, their space differed greatly from ours. The distance of light-year may be just proximity in another space. The UFO only needs to enter another space and flights a short distance and then, when they break through the space back, they are away from many light years. Therefore, the aliens can easily do interstellar travel to our planet without too much time. We humans are trapped by the molecular space and use low-level aircrafts. Even if the pilot gets old, he cannot fly at much distance.

The flight in other space is really "coming and going without a trace" due to differences in time and space. When a man witnesses a UFO, sometimes the flying saucer is suddenly gone. It is because it goes to another space so it is invisible. When it returns to this space, he will feel it suddenly appear in the air

and that is very magical.

Albert Einstein once said: "The world is basically unknown. Currently, the understanding of humans on this world adds up to not more than one percent by wisdom." Some people may not believe in the existence of another time and space. Let me prove the existence of another space with a simple theory.

Which space do humans live in? As we all know, the biggest object we see is a planet. The eight planets comprise the solar system and many galaxies comprise the Milky Way. The smallest object we are exposed to is molecular. The particles smaller than the molecular can be observed and functioned only with the help of instruments. We humans live between the molecules and the planet, which are the space we stay.

But think about that the molecule is composed of atoms and is there any space inside the molecule? Atom is composed of nuclei (protons and neutrons) and electrons. Is there any space inside the atom?

We know there is space inside the molecule and the atom through the instrument but we cannot see it without the instrument. We also cannot function the molecule and the atom simply by hands.

Trapped between the molecules and the planet, we humans cannot see the additional space, not to mention function in the additional space.

The internal space of the molecule is very small, but when the interior spaces of all the molecules in the universe are added up, they will be a vast space! Similarly, although the interior space of an atom is very small, when the interior spaces of all the atoms are added up, they will be an incredibly big space.

Let me express the above theory with a simple example. It is supposed there is a mountain in front of us occupying the space; if we do not remove the mountain, we could not enter the space. At this point, we imagine our bodies are getting smaller and smaller to become a little life on an electron inside the atom. Look at that mountain now and you will find the scene completely changed. The mountain has become a huge galaxy and countless planets (in fact, they are protons, neutrons, and electrons) in regular running. If we could fly, we would have to fly into the interior of the mountain. Wow! Within the mountains is an extremely vast space. The countless electrons are the planets; perhaps there are humans on these planets. What a wonderful world! At this point, our bodies gradually become larger. Bad! We are going to bump the "planet" and we must exit the mountains. We come to the mountains outside to look at the mountains. The mountain is a huge universe; internal stars run and life is prosperous. We can not only see but also function it.

Let us expand the vision to look around the mountains. The surrounding air originally cannot be seen with the naked eye. At this point, all the space of the atoms inside the air molecules is vivid. It is also a galaxy. The galaxy density here is much smaller than that within the mountains, showing the status of a free flowing.

In fact, we have entered the additional space. The maximum objects in the space are the "planets"-protons, neutrons, and electrons, and the smallest objects compose particles-quarks. (In the modern scientific hypothesis, electrons, neutrons, and protons are composed of more fundamental particles-quarks.) We have entered the space located between the quarks and electrons. You will find that it is broader than our space large to infinity of the universe. There are also prosperous lives and infinite vitality. In this space, you can see and touch them.

The ideological journey ends and we go back to our space. All the wonders are gone; we cannot see inside the mountains nor enter the interior of the mountain. The space is between the planet and the molecule, between the molecule and the atom, between the atom and electrons, neutrons, and protons, and between electrons, neutrons, protons and quarks. The matter is infinitely divisible and the space between the particles is also infinitely countless. This is the reasoning to the microscopic direction. On the contrary, to the macroscope, the reason is the same. Different space has a different distance concept. The more microscopic the space, the more broad the scope of the internal space. A mountain in our space becomes a huge galaxy in the space between quarks and electrons. Of course, this is only because of relatively larger dimensions to see the electrons as the "plane" so the spatial extent is relatively larger.

Different space has a different concept of time. Time is actually determined by the decay rate of the material. The faster the rate of decay, the faster the time. The slower the decay speed, the slower the time. The decay of molecules and quarks is at different rates and the reflected two space-times are not the same. Different space has different substance appearances.

Water in this space, for example, is the form at the molecular level.

If we can see the morphology of water appearing on the atomic level, i.e. see permutations morphology of a hydrogen atom and an oxygen atom, what kind of morphology is that? We can imagine that must be a very different morphology. Aliens made their base in the Gulf of California. In this space there is the ocean. However, if we break through this space into another space, there is not sea water but a valley people can live. I could not help but sigh, this universe is really fantastic!

In terms of the fire, for example, the sun is a huge fireball in our space. But think about it, can the fire of the sun burn the atom? Not, the sun fire is not the real fire and it cannot burn other space, only burning between the molecule and the planet we live in. When you break through this space into another space, you will find that the sun turns out to be a cool world and there is no trace of the hot. The god Apollo rules the planet. Similarly, the other planets of the solar system have life to live. Mars, Venus, and Saturn people are real but exist in another space. To expand, all objects in the universe have different appearing morphology in the additional space.

It is as if all the space existed at the same time. It seems to coincide but actually each occupies a layer of space without interfere with each other. We humans only occupy the molecular space but there is no interference with

the atom space and the quark space. Likewise, the other space does not interfere our space. The universe we see is just the showing cosmic form in the molecular layer, not the whole picture of the universe. All space together is a real show of the universe.

The above theory may be more abstract. Let me cite a simple example to facilitate understanding.

"Bermuda Triangle" is a triangular devil area located in the Atlantic Ocean. A lot of vessels and aircrafts enigmatically disappeared there. Maybe after a few years, ten years, or decades, they miraculously appeared. In fact, there is a marine area leading to another space. When vessels and aircrafts came there, if unfortunately another space's "door" just opened, they were easily to be into there. Tt is not a formal one door but a state of being accidentally.

Missing persons often said strange words. The U.S. military recorded the words said by the missing person that "Huh? How will this look like it?" It was actually the missing person entered another space and saw the substances different from ours or experienced a completely different psychological feeling from our space.

After many years, the missing persons might return to our space, not necessarily in the crash area but suddenly appearing in another sea. Because different space had a different view of the distance, they felt they were still in the original waters but actually they came to another waters. Their family would have asked them curiously what they did for so many years in missing. They felt they just swung in the sea for a few hours. Because different space has e a different concept of time, in staying in another space for a few hours, many years have passed in our space.

The above description is only a theoretically simple proof of the existence of another space, roughly about a few features of the other spaces. Aliens and the people under the ground have long been breakthrough another space in practice and conducted space travel with the features of another space. The way they develop their technology is not the same with ours. Our descendants will break another time and space to prove that what I said is true.

Section 30
Magic Way and Shura Way/ Positive Aliens and Negative Aliens

We call the non-malicious aliens as the positive aliens, such as the Pleiades galaxy people and the Andromeda galaxy people. They instruct humans how to enhance mental awareness. The malicious aliens, such as lizard people and gray people, are called as negative aliens. They would kidnap people and injure humans.

Now a lot of people have this doubt that since the aliens exist, why do they not contact humans formally nor occupy the Earth? That is also the reason why a lot of people say that there is no alien. This time, I will tell you the mystery. Many people believe that the occupation of Earth is not an easy thing for aliens with their technology level. There are a lot of people thinking positive aliens prevent the conspiracy of the negative alien, so aliens failed to capture the Earth. Is that real? Then I will first talk about the positive aliens. Are positive aliens stronger than negative aliens? In past galaxy war history, positive aliens often fail to fight against negative aliens. It may be doubted that even if they are powerful, why they didn't appear on Earth to prevent negative aliens from destroying? Will people on earth be more grateful and regard them as gods? Do they do goodness without seeking recognition? Some people say that they are for the purpose that not interfering human beings since the advanced race does not interfere with the development of low-level ethnic civilization. That being the case, the alien races are not at the same level. The senior races often guide the development of civilization of low-level race. Why don't they do so for people on Earth? Imagine when a civilization has interplanetary navigation capability to explore the universe, will they give up exploration for there is life on that planet? Is that not the goal they explore? When aliens see there is also low-leveled intelligent life on other planet, are they afraid of contacting them? Aliens have researched mankind for thousands of years. They are not afraid of humans.

It was also said that because human beings learn to use nuclear weapons, aliens worry about the destruction behavior of humans harming their survival, so they warn humans in extensively appearing. Even if the Earth is destroyed, it would affect the solar system at most, not other galaxies. So why are the aliens in many other distant galaxies concerned about the Earth? Others say that's because they could not bear the destruction of the human species, so when the humans encounter doomsday crisis, they would come for rescue. However, there are numerous planets in the universe and each planet actually has life with different ways of life to survive, living environment, and living space. Destruction and birth of a planet occurs in every moment so that even if the planet is destroyed and life disappears, are aliens surprised at that? Why do the positive aliens not save those lives, instead being so interested in the people on Earth?

Let's talk about negative aliens. If positive aliens really have so many causes to be not in contact with humans, negative aliens will not have these problems. Their purpose is aggres-

sion to seize the planet's resources. It past Galaxy's history, they invaded the planet many times. Did they make actions sneakily? Even if there was positive aliens' obstruction, negative aliens could confront with positive aliens on the Earth or the Earth's solar system. Actually, the wars between positive aliens and negative aliens are broken out on the edge of the solar system; some were once large-scale war, but we human beings could not directly observe them. Is it necessary for them so sneaky?

Some people say it was a cosmic war. Aliens' warships were enormous and the solar system was too small so they fought outside the solar system. Indeed, some of the alien motherships are more enormous several times larger than the Earth. The solar system is indeed a very small place for their wars. There are also the bases of negative aliens and positive aliens on Earth. Their countless small aircrafts shuttle freely above the Earth. Why did we not see them at war in the air? Why are both positive and negative aliens afraid to openly and formally contact with mankind? Don't they want to occupy the Earth? Wrong. In fact, they are dreaming of possession of the earth and ruling mankind since ancient times. Before the human civilization, they noticed humans. Although they are racking their brains and intriguing for the occupation of Earth, possessing human bodies, and ruling humans, it is not really an easy task for aliens to achieve this goal.

Aliens had been contacted and studied the Earth people since the ancient times. Many ancient books recorded the aliens-related history. Aliens had long found the human body was perfect and they could not create humans with heir level of science and technology. Unfortunately, human beings do not know the mystery of us but long for the alien body, such as the robot. The aliens attempt to occupy the human body, but defeated by human. What an incredible thing. Although alien technology is quite advanced, it is still a basic knowledge for some people. These people are the cultivators believing the Buddhism God and having supernatural powers. The alien term is our modern name. For ancient people, especially the ancient Chinese people, aliens were no strange for them.

In ancient China, Taoism often mentioned ghosts and goblins; aliens belong to ghosts and goblins. Buddhists often mentioned six roads of reincarnation, in which there is Shura Road and aliens belonging to it. The organisms in Shura Road were the living creature looking like humans but not having human nature. They were devils and regarded as beasts. They like to eat people. People always think devils look terrible, are bloodthirsty, and may injure humans. It's completely wrong. Such devils are just a part of the devils, called demons. Satan is the leader of demons. Although aliens belong to devils, they vary and not all may harm humanity. A lot of devils look beautiful. Some have no enmity to humans. Some help humans for some purposes, such as being possessed on human body for treating diseases and warding off evil. Some also concern about kindness, teaching how to improve the love of mankind and how to enhance.

Some say it is very good and they are similar to humans. Not the same. We call them devils not because they are evil but they do not have the human nature. Life has its good and evil sides. Even the devil is also friendly to its similar group. Not to say that only the one regarding goodness is human. Monsters also regard goodness but they are not humans.

The ancient people often mention yellow foxes and white willow and demons s would possess the human body to absorb the essence of people and practice to the human form. Alien are also the possessing body. Who learns them will become the devil and his life will be controlled by the devil. Do you want to become like this? So, in ancient times, whether Taoist or Buddhist cultivators must eradicate the possessing bodies which were harmful to humans. Since ancient times, we heard of killing demons and surrendering monsters, which referred to such things.

Every year, many people are captured by aliens; some are researched, some are eaten, and some are taken to the outer planets as animals in the menagerie.

The alien belongs to the devil and the devil has magic powers. Whether it is from his forces or with the other tools, it is supernatural power. Of course, it is easy to deal with ordinary people. But for the cultivators believing Buddhism God, the devils are early to be defeated. The supernatural war in the myths and legends of ancient times is the war of cultivators and aliens. Because the majority of the ancient people believed in God, they were protected by God. It was the cultivators believing in Buddhist and the God in charge of humans protect humans from being invaded by aliens. Human body is like the Tang monk for aliens. The monsters which wanted to eat Tang monk were bound to render evil and finally they could not eat it and lost their lives.

Section 31
The Aliens Counterfeiting God-the five major steps in the occupation of Earth

Although mankind have God's protection, are those evil aliens willing to give up occupying mankind? Of course not. So how do they do? They conducted careful arrangements.

First, they cut off the contact between humans and Gods, letting humans give up the belief in God and thus God cannot protect humans. For several thousand years, the contact between humans and God has been through religious defenders. So, to occupy humans, the first step is to make people no longer believe in a religion. Humans' religious beliefs maintain thousands of years. Until modern times, when people became more and more complex and morality gradually corrupted, more and more people no longer believed in God. So, the aliens began to make actions. Atheism system's masterpieces, "Heliocentric Theory" and "Theory of Evolution" appeared. Originally, these were just hypotheses. But, they were consistent with people's mentality of being arrogant and attempting to conquer nature of the world. So, people regarded truth and faith as hypotheses, easily discarding religious and god beliefs for thousand years. This result lets the aliens very happy and they did not expect the first step was so easily to be completed.

Here I will not explain why I think "Theory of Evolution" was not truth since I'd mentioned it before. Let me explain the "Heliocentric Theory." Religious theory is that the Earth is the center of the universe and the universe turns around the Earth. In the theory of the "Heliocentric Theory," the earth turns around the sun and the phenomena people observe now confirmed this point. So People recognized the "Heliocentric Theory," abandoning the religious theory of the "geocentric theory." Where did the religious theory of the "geocentric theory" originate from? It was from God telling the mankind, which is said in terms of the range of the whole universe. In the "Heliocentric Theory," it was considered in the range of the solar system. In fact, both are correct and not contradictory. People relying on scientific tools do not see the phenomenon that the Earth is the center of the universe. However, the thing that you cannot see it does not mean others do not also see it. The cultivators reaching a high level can see it but they cannot let other people see what they saw. Therefore, cultivators firmly believe in religions and God. The ordinary people can only observe the situation the Earth rotates around the sun and thus deny the "geocentric theory."

Over the past middle Ages, Christianity dominated the Western human society. It was the theocratic form of rule. The contradictions of the "Heliocentric Theory" and the religious theory affected the religious ruling base. In order to safeguard their own rule, the people of Christianity burned Bruno who insisted in the "Heliocentric Theory" to death. This was also the blemishes on the history of religion, which is a very disgraceful, very unpopular thing. Because the theocratic form of rule caused many tragic events and contradictions, people began to doubt the religion and finally abandoned the religion as well as their faith in God.

Most people do not get one thing that religion was founded by humans, instead of God. God gave people the truth of the universe and how to practice to go to senior heavenly world. God simply does not recognize religions or theocracy. The decline of religion does not mean the decline of God. Religious clergy corruption is not equal to God's corruption. However, because human faith has associating relations, distrust of individual behavior became the distrust of the personal representative religion, eventually turning into the distrust of religious representative God. As a result, humans abandoned belief in God as well as the traditional culture God sent to humans but believe in the believing-based empirical science which is in fact the alien science.

Second, the aliens installed technology and ideas to cultivate humans into aliens. In fact, the empirical science is also a religion. Traditional human culture and alien culture are completely different. Traditional culture and religion focus on spiritual development, from the lower spiritual level to the higher level of substance. Extraterrestrial modern empirical science and culture are exactly the opposite, focusing on the material level, from the low-level material aspects to a higher spiritual level. Therefore, this form of religion is not easy to be noticed.

For the ancients who believed in God, the traditional methods could more deeply study the human body and the mysteries of the universe. But for modern people who do not believe in God, this method is completely useless and cannot achieve even the most shallow discovery range. If you do not believe in God, a field is generated around your human body like a diaphragm, automatically isolating from the characteristics of the universe so that humans can not perceive a higher level of cosmic phenomena.

For people who do not believe in God, alien culture fits their needs of discovering the world. Aliens instilled alien technology system to mankind systematically since the Industrial Revolution. This technology, developing to today, is now the western empirical scientific system. Now what people learn is actually an alien culture. Any substance has its own material form of existence in another space. Information is the same. When you see something, it forms a layer of material in the field of your space. Over time, it will form a layer of body. Humanity learns alien culture systematically since childhood, so the additional space on the human body has a layer of the alien body. The layer of the alien body also has thoughts. They are alive and can manipulate the people's minds. Many Western inventors have a variety of inspirations and ideas emerged so that a variety of new machines are invented. Where are these inspirations from? They are not thought out by the inventors but from the installation of the layer of the alien body to instill alien world's information to humans.

It may be doubted that why God do not stop it? It is because humans want it so God will not intervene it. What is the purpose of the aliens to do so? For one thing, they want people to reproduce the alien world on Earth. When the alien world is completely reproduced on Earth, it will be the time the aliens completely rule the humans. For the other, if the aliens occupy the Earth forcibly and install alien technology,

then God will destroy them. However, when humans initiatively believe in alien culture, God will not turn a blind eye. When most people believe in alien technology, God will no longer care about humans since people believe in devils so that humans s completely cut off contact with God.

Although the aliens manufactured a layer of extraterrestrial body on the human body on to control people to some extent, this is not the possession of the body, just like being possessed. Once the people re-believe in the positive god and practice Dhamma, the God will help people to remove the possessed layer of the alien body, which is easy for the positive god. In this way, the aliens could no longer control people and their plot would spoil.

The aliens found that the way of possession to rule the humans was not the best, some humans were the cultivators believing the Buddhist god and the aliens were unable to control them or fight with them. However, even the cultivators were powerful, they could not use extraordinary power to publicly interference the development of human society. The aliens discovered this, so they planned a new conspiracy.

Third, the aliens let humans to create alien life to replace human beings. We humans are of the carbon-based life form. A lot of people have heard of the description that aliens are of the silicon-based life form. What is the silicon-based life form? It is computers and robots. The main component of the computer is Silicon. The robot with self-thinking skills is the silicon-based life.

Since 1980s, computer technology has been developed rapidly in the globe. Advanced chip is constantly updated and today's society is the age of universal access to computers. On all the people who can operate a computer, aliens made a layer of alien body, except only of Dhamma cultivators. The ultimate development goal of the computer is the artificial intelligence robot in the combination of computer and mechanism with independent thinking ability. Therefore, in order to achieve this goal quickly, under the control of the alien, the robot technology in the past period of time suddenly leaped.

Now, Transformers, robots, aliens, and related cartoons were created to install humans and their cultures were from the installation of aliens to let humans in assimilation of alien technology for the purpose that the intelligent robots can completely replace humans eventually.

However, while the priest climbs a post, the devil climbs ten. (As an aside, the statement was the devil installing to humans. Devils are never higher than the road since their levels are too different.) Aliens' conspiracy did not escape from the eyes of God. So God reminds people to be careful of aliens' conspiracy in the human culture, such as in ID4 movie series. People now have a full understanding on aliens' conspiracy and are very cautious for the development of artificial intelligence. Robotics development came to a halt compared to the past. At this point, aliens' conspiracy of replacing humans with robots failed. However, aliens did not want to stop because they'd not like to waste the careful preparation for hundreds of years. So they have a new conspiracy.

Fourth, the aliens let humans to manufacture cloning to replace humans to destroy humanity finally. To become a normal person, a man needs not only human flesh but also the primordial spirit (which is similar to the soul but not identical), temper, mettle, and other factors, which are human nature. Otherwise, humans are just the flesh without ideological activities.

The birth of normal human is completed through the binding of male and female sexual reproduction cells so that the life is given with the primordial spirit and other factors and the life can be lived. Other sexual reproduction animals do the same, too. For the non-sexual reproduction cloning organisms, whether humans or animal, God does not bestow soul. If human cloning is manufactured but God does not bestow him soul, what will enter the body? It's the aliens. They enter the human body to dominate human flesh and obtain the Tang Monk they dream of.

Although the cloning flesh is the same as the parent, the primordial spirits are completely different. Although clonings might look like humans, they are devils in the eyes of God. God does not admit that he is human because he has no human nature.

Why atheists always publicize people have no soul and wen people die, their thoughts are also dead is because aliens did not let humans find the secret. Person's ideas are not derived from the flesh, but from the primordial spirit, namely the soul. When people die, it is just the death of the flesh cells composed of molecules. However, the smaller particles such as atoms comprising the primordial spirit will not die. It is just like being off a layer of shell. In the next reincarnation, a shell with another molecules are composed, which may be the human body, animals, plants, or stones ... etc.

The molecule is constituted by the smaller particles such as atoms. If molecular perish means everything disappearing, the smaller particles such as atoms constituting molecules should be also destroyed. When atomic destruction occurs, there will be nuclear explosion. Similarly, if people's death means everything disappearing, it is to say all material particles constituting the human body should be destroyed so that nuclear explosions occur. Have you ever seen nuclear explosion occurs when people die? I guess not! So isn't this argument of atheism wrong?

If the clonings of extraterrestrial life are allowed to exist, they will certainly replace the traditional fertility because there is no pain of human fertility. Then when there are more and more clonings, they will enact legislation to require human cloning rights, finally completely banning the traditional way of fertility. Then, the real human beings really completely are destructed and the alien devils will completely replace humans to rule the world.

However, the conspiracy is not escaped from the discernment of God. Although cloning technology has been rapidly developed in recent years and there have been constant requirements for the legitimizing of clonings, this problem is soon attracted the attention of all mankind. Under the pressure of traditional concepts and religious ideas for thousands of years, all countries in the world have reached agreement to legally not legalize cloning test for reproductive purposes. In fact, this is God's arrangement, again completely destroying aliens' conspiracy.

Because aliens cannot see the existence of God, they dare to trap humans. In the universe, however, there is a causal relationship. No matter what you do, there is retribution. What goes around, comes around. Because aliens' repeated action of conspiracy interfered the arrangements of God and the positive believe of humans in God, those who participate in the conspiracy aliens and their planet will be wiped, in which many have been wiped out. Some evil aliens slipping through the net fled to the Earth, disguised as people, corrupted human morality, and destroyed traditional human culture, but Ordinary people simply could not tell them. In the last trial, God will wipe them out.

The aliens who will perish already know the results, so they want to harm people to die. So, the new conspiracy began.

Fifth, aliens pretend God to let people think they are the creators and God so as to accept their beliefs. Then, humans may be ruled by them and destroyed as their grave goods.

God and aliens are completely different. Aliens and humans are at the same level, a different space, in the lower three realms of space. However, God is in the high level of the three realms, in the heaven beyond the three realms. The level and capacity of both differs a lot.

In previous several conspiracies, alien secretly manipulated humans and they did not show up. Also, it is also impossible to detect it is related to the aliens. Now, they personally perform their conspiracy.

In recent years, they began to appear in large numbers around the world, letting people observe them and see their state-of-the-art scientific and technological level. Some of them contact humans directly and others contact humans through telepathy, such as the Western psychics.

Aliens control them and instill false information, saying aliens are God and created humans. Besides, aliens said the gods or sages people know, such as Jesus, the Egyptian sun god, China's Lao Tzu, India Buddha, Middle East's Mohammed…etc. were all alien heralds, and aliens will personally send spacecraft to save humanity to other planets in the end, aliens will instruct humans' upgrading and evolution to a high dimension. In short, they used a variety of means to let people believe they are God, the creator of mankind, and the savior.

In many ancient myths, it was mentioned different gods created different nationalities according to his own image. Man is God's creation. If people deny their own gods creating them, it is equal to alienating way out and God will eventually ruin mankind.

Aliens know this, so they let people blindly believe modern science and atheism to let people believe and choose them and God will not be able to control everything. Because there is free will principle in the universe which is selected by humans, God not only does not stop aliens but destroy the humans following aliens. If one chose to follow them, meaning to be the same kind of them, then he will be destroyed when the races are destroyed eventually. It is equal to self-defeating of humans and aliens' conspiracy succeeds.

Imagine if aliens made humans, then why there is human only on Earth? There should be people everywhere in the universe. In the ancient myths, it was mentioned that different gods created different nationalities according to his own image. If aliens are those gods, why don't they look like us?

Indeed, some aliens look almost like people on earth, and it is caused by other reasons. It was because their ancestors were created by God. God did not only make humans. Their ancestors were the test products when God created the perfect life forms of humans, namely the semi-finished products. They were also made on earth in the hundreds of millions of years ago. Their ancestors could be regarded as the people on earth, so they would be closer to people. Although they look like people and some are almost the same wit people, their structure is not the same with us. Ordinary people cannot see the difference, but cultivators can.

These aliens looking like people are not made in the same period. Every time when God made the people, god lets people to develop civilization, which was the source of prehistoric civilization. After a long test, the one who was not perfect, he would be eliminated. Each god retained a small portion of people, sending to other planets to retain their advanced civilization and let them continue to multiply. They were the witness of the creation in the universe development process. These life forms were the ancestors looking like aliens. The alien life was longer than the people on earth in thousands, tens of thousands of years, the longest. After hundreds of millions of years and the replacement of countless generations, their descendants have forgotten the existence of God and regard it as a legend.

In fact, even aliens do not know these secret themselves. When some aliens had the ability to see things in a certain level in another time and space occurring in the future, they would tell humans what the future would be. Now there are many prophecies about doomsday disasters from aliens on the web. In them, there are some things about aliens. For example, the UFO would appear in a city, which might be accurate news. Other prophecies like the doomsday on Earth are not true. Why is that? The aliens cannot imagine why things are not always the same. The answer is actually very simple. Those disasters were controlled by God. Different gods control different disasters. If alien prophecy has a high accuracy, many human beings may believe aliens, instead of God, that alien conspiracy would succeed. God, of course, does not allow it. The gods would change the disaster arrangements for the future so that the alien prophecy became inaccurate so that people naturally would not believe that aliens.

It was evil, negative aliens involved in the conspiracy. Not all aliens were involved in the conspiracy. Some positive aliens were well-intentioned for the salvation of the human race, but their capacity was limited. They could only see the lower space and they did not know the truth of the universe large purifying changes.

They knew their own space was about to be destroyed and as long as they enhanced the levels and evolved to the high-level space, they could escape from disasters. But they did not know the universe purification change was in all space from the top to the bottom, without exception. The substances and life forms which did not meet the new universe standard would be destroyed and eliminated.

Section 32
Universe Purification and the Aliens Fled to the Solar System

Purification had been completed outside space in the Milky Way a few years ago. In recent years, astronomers have observed and found the spectacle of the destruction of many old galaxies and the birth of new galaxies and their number are difficultly to imagine. In fact, these are the real show of the purification process of the universe in human space.

Now, the scientists also found that the universe is in the rapid expansion; the Milky Way, including the solar system we live in, is rapidly away from other extragalactic systems. If this continues, the Milky Way will be let alone in the universe and scientists currently cannot explain this phenomenon. In fact, this is the real performance of the universe during the final purification process. Now, just the Milky Way has not purified. So God has isolated the universes which have been purified from the Milky Way, otherwise the Milky Way would pollute the others.

The space humans and aliens live in is in the different space of the universe at the lowest levels. Purification has now been completed in the spaces except for the Milky Way. The planets and space aliens survive have been purified. The aliens involved in conspiracy have been destroyed by God and the rest have fled to the Milky Way.

There are a lot of spaces in the Milky Way from the top to the bottom. Our space is in the underlying space of the Milky Way and many aliens hide in different spaces here. Now the high-level space within the Milky Way is purified and transformed into the new universe. The planets aliens live have been destroyed. The aliens are forced to flee to the lowest space in order to save life, which is humans' space. So there are many aliens disguising as humans in human society.

Because aliens cannot see the presence of God, they think the purification of planet destruction is just a natural phenomenon and they can survive as long as they flee to another space.

Currently, the purification trend of the universe has reached the living space of underlying humans and aliens in the final stage of the purification. It can be said that the purification process of the universe is now in the final stage. But the present means of observation of human technology is far behind so people cannot be aware of the eliminations in the universe at all levels of thrilling.

The purification process of the universe is the process of transforming old universe into the new universe. In human space, it is the human purification process. God arranges it from 1992 to 2012, in a total of 20 years. This is the origin of human purification and the Mayan doomsday prophecy the Mayan mentioned. They predicted that God deliberately left them to alert the eschatological people of today.

God arranged a finale for the purification process in the human space, which is the Gods' doomsday trial mentioned in the Bible. After the trial of the gods, human space purification is completed, marking the entire universe is purified and a new birth of the universe.

Some aliens have a certain ability to foresee the future, but they foresaw the originally arranged purification process of the old universe. Now the God has changed the purification process, so they cannot see it.

The situation that aliens' living space will be destroyed expected by aliens is the showing of the space universe purification process and aliens are the objects to be eliminated in the purification of the universe because all the aliens do not comply with the new standard of the universe.

Imagine that aliens do not meet even the new standard of the lowest levels of the universe now, not to mention the high-level standard. If they do not know the high-level standard, how do they get evolution to a high level of space? Even some aliens have the ability to open the high-level space, they cannot survive. Senior space already reformed to reach new universe standards so the materials and life forms are not allowed to exist. As long as the aliens cannot comply with the new standard of the universe, they would be the objects to be eliminated no matter what level they flee to. Unfortunately, the aliens cannot see the senior space so they are not aware of this.

The alien life will be all-out after the Doomsday. The aliens joining the conspiracy will be destructed and those which not involved in the conspiracy will be converted to other creatures in the new universe after reincarnation. Because the modern science aliens passed down to humans is not suitable for human development, it will also be eliminated and there will be new culture to replace it after the New World.

Why human bodies are so precious?

The human body hasn't been existed in the universe in the past and it is the most exquisite life form since God created the universe. After countless years of experience, human bodies have become perfect. The human body can shuttle at any time and space of the universe freely through practicing and it was impossible for even God in the past.

Now, any God wanting to obtain the body must enter the Three Realms reincarnation to be humans. Their all memory must be erased. In this state, God became a man and might get lost in this world until the destruction. In the past, no one of all the gods entering the three realms could return to the Three Realms.

Although Tang Monk was good, eating Tang Monk came at a price, even the price of life. In the past, once a god entered the three realms, in the eyes of other gods, this god was dead, so no one dared to enter the three realms.

Aliens foresaw the world they survive will be destroyed. They knew the human body can transcend space and they could escape from the situation of destruction by having the human body. So they did everything to get the human body. However, they did not want to be reincarnated humans so they used dishonest methods to get the human body.

Unfortunately, humans now have the human body but they do not cherish it. What a sad thing!

Section 33
How Do Humans Survive the Crisis of the Universe Purification

Although I write the myth novel, the above-mentioned is definitely not a myth. I firmly oppose to human cloning. It is the bottom line of human life and death and I hope every reader could have such wisdom and foresight.

The people under ground were closed in the earth by aliens and lost contact with the moon. The situation was very bad and they had been trying to change this situation. Decades ago, a king under the ground pulled out the Excalibur so he was mandated to the world to search for the way to save nations and the moon. He ran out of the ground and then killed by aliens. The primordial spirit reincarnated to China and fortunately practiced the Law of Creator.

In fact, after the start of this period of human civilization, the moon defenders (In addition to Yi and Chang-Er, there were many other kings practicing to the moon) have reincarnation to China and practice in the Law of Creator (Dafa). With the joining of the cultivators, the positive powers were greater and greater. In this case, the positive gods in the universe were gradually cleaning up aliens.

Mayan prophecy tells us humanity has entered an important period-earth purification period, from 1992 to 2012. During this period, all things on earth should be purified and all corrupt life and everything would be eliminated, including all variations, distorted, dirty, and evil factors. Every cell on Earth, including humans and aliens, will go through this process. This is the law of survival of cosmic objects, and it is irresistible. Aliens did a lot of bad things on Earth so they are eliminated by the positive gods. The bad guys in this period are also in danger. If they do not learn from bitter experience and cultivate with goodness, they will also be eliminated. The good guys have to do better, be wise and cultivate the mind to get the protection and bless of God, smoothly through the purification period.

After December 21, 2012, Earth purification ended and mankind will enter a new era. The ancient Mayans predict the irresistible luck and left prophecy to enlighten those who can witness this great moment in history. The people who are lucky to enter a new era are blessed good ones. I continue to look back, seeing the next scene, but I cannot reveal it. Here I can only advise people of doing goodness. What goes around, comes around. The people with good deeds will receive good fruits. The people with all kinds of evil must be punished, which is eternal and unchanging truth!

It's time to end the cosmic voyage. Now, it's night. I look out of the window and there is a bright moon in the sky.

Postscript

"2012 Truth Trilogy" includes "Ancient Myth", "The King of Kings", "War between Good and Evil".

In the second part "The King of Kings", the story focuses on the savior prophecy and the truth of the legends waited by the people in the East and the West.

In the messages we have explored, the saint whom the people in the East and the West are waiting for may be the same person.

Is there any relationship between the "savior", "Christ," "Messiah," the King of Kings "and the "Zi-Wei saints", the Buddhist's future Buddha "Maitreya"?

Does he have the ability to rescue all the problems of modern humans?

This should be the issue that each one of Earth's humans wants to know the most.

And you will be the first people to see the truth.

Please look for the upcoming issue.

The King of Kings of 2012 Truth Trilogy

Ancient Myth

Series Name: 2012 Truth Trilogy
Volume in Series: 1
Author: MicroStar
Translator: Vicky Hsieh
Publisher: ZUXAR, LLC.
Publication Date: August 17, 2015
zuxarinfo@gmail.com
ISBN-13:978-1516933549
ISBN-10:1516933540